2

AMERICAN HISTORY, CULTURE AND LITERATURE

WONDERS OF THE YELLOWSTONE

AMERICAN HISTORY, CULTURE AND LITERATURE

Additional books and e-books in this series can be found on Nova's website under the Series tab.

AMERICAN HISTORY, CULTURE AND LITERATURE

WONDERS OF THE YELLOWSTONE

JAMES RICHARDSON

snova
New York

Library of Congress Cataloging-in-Publication Data

ISBN: 978-1-53615-079-7

Published by Nova Science Publishers, Inc. † New York

CONTENTS

PREFACE*

The first earnest attempt to explore the valley of the upper Yellowstone was made in 1859, by Colonel Raynolds, of the Corps of Engineers. His expedition passed entirely around the Yellowstone basin, but could not penetrate it. Ten years after Colonel Raynolds's unsuccessful attempt to solve the problem of the Yellowstone, a small party under Messrs. Cook and Folsom ascended the river to the lake, and crossed over the divide into the Geyser Basin of the Madison. The general public were indebted for their first knowledge of the marvels of this region to an expedition organized in the summer of 1870 by some of the officials and leading citizens of Montana. In the meantime, a large and thoroughly-organized scientific party, under Dr. F. V. Hayden, U. S. geologist, were making a systematic survey of the region traversed by Colonel Barlow. It is safe to say that no exploring expedition ever had a more interesting field of investigation, or ever studied so many grand, curious and wonderful aspects of nature in so short a time.

---

* This is an edited, reformatted and augmented version of "Wonders of the Yellowstone" by James Richardson, originally published by Scribner, Armstrong and Company, dated 1873.

Chapter 1

THE CROWN OF THE CONTINENT

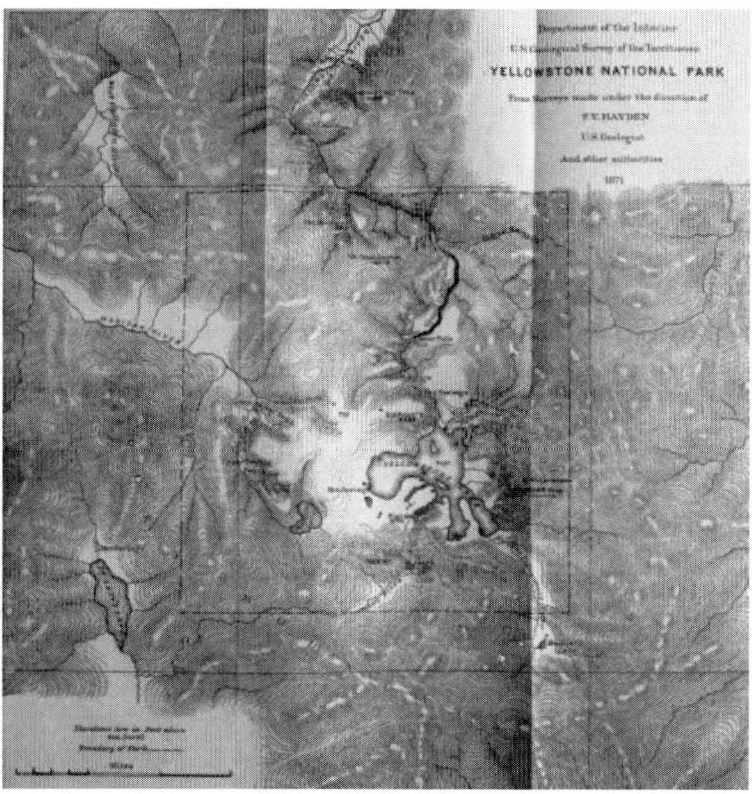

Hayden's Geological Survey of Yellowstone National Park.

In the northwest corner of the Territory of Wyoming, about half way between the Mississippi River and the Pacific Ocean, and in the same latitude as the State of New York, the grand Rocky Mountain system culminates in a knot of peaks and ranges enclosing the most remarkable lake basin in the world. From this point radiate the chief mountain ranges, and three of the longest rivers of the Continent—the Missouri, the Columbia, and the Colorado.

On the south are the Wind River Mountains, a snow-clad barrier which no white man has ever crossed. On the east is the Snowy Mountain Range, and the grand cluster of volcanic peaks between it and Yellowstone Lake. On the west is the main divide of the Rocky Mountains. On the north are the bold peaks of the Gallatin Range, and the parallel ridges which give a northward direction to all the great tributaries of the Missouri from this region.

Set like a gem in the centre of this snow-rimmed crown of the continent, is the loveliest body of fresh water on the globe, its dark-blue surface at an elevation greater than that of the highest clouds that fleck the azure sky of a summer's day, over the tops of the loftiest mountains of the East. Its waters teem with trout, and the primeval forests that cover the surrounding country are crowded with game. But these are the least of its attractions. It is the wildness and grandeur of the enclosing mountain scenery, and still more the curious, beautiful, wonderful and stupendous natural phenomena which characterize the region, that have raised it to sudden fame, and caused it to be set apart by our national government as a grand national play-ground and museum of unparalleled, indeed incomparable, marvels, free to all men for all time.

Evidences of ancient volcanic action on the grandest scale are so abundant and striking throughout the lake basin that it has been looked upon as the remains of a mammoth crater, forty miles across. It seems, however, to have been rather the focus of a multitude of craters. "It is probable," says the United States geologist, Dr. Hayden, with his usual caution, "that during the Pliocene period the entire country drained by the sources of the Yellowstone and the Columbia was the scene of volcanic activity as great as that of any portion of the globe. It might be called one vast crater, made up of a thousand smaller volcanic vents and fissures, out of which the fluid interior of the earth, fragments of rock and volcanic dust, were poured in unlimited quantities. Hundreds of the nuclei or cones of these volcanic vents are now remaining, some of them rising to a height of 10,000 to 11,000 feet above the sea. Mounts Doane, Longford, Stevenson, and more than a hundred other peaks, may be seen from any high point on either side of the basin, each of which formed a centre of effusion."

All that is left of the terrific forces which threw up these lofty mountains and elevated the entire region to its present altitude, now finds issue in occasional earthquake shocks, and in the innumerable hot springs and geysers, whose description makes up so large a portion of this book of wonders. Nowhere else in the world can the last-named phenomena be witnessed on so grand a scale, in such limitless variety, or amid scenes so marvellous in beauty, so wild and unearthly in savage grandeur, so fascinating in all that awes or attracts the lover of the curious, the wonderful, the magnificent in nature.

Chapter 2

First Explorations

In their exploration of the headwaters of the Missouri in the summer of 1805, the heroic Captains Lewis and Clarke discovered and named the three terminal branches of that river—the Jefferson, the Madison, and the Gallatin; then ascending the first named to its springs among the Rocky Mountains, they crossed the lofty ridge of the divide and pursued their investigations along the Columbia to the sea. The following summer they returned, separately exploring the two main branches of the Great River of the Northwest, each perpetuating the name and fame of his brother explorer by calling a river after him. Ascending the southern, or Lewis Fork, Captain Clarke recrossed the mountains to Wisdom River, (a branch of the Jefferson,) then traversed the country of the Jefferson, the Madison and the Gallatin to the Rochejaune, or Yellowstone, which he followed to its junction with the Missouri, where he rejoined Captain Lewis. The map of the country explored by these brave men, makes the source of the Yellowstone a large lake, doubtless from information received from the Indians, but they seem to have heard nothing of the marvels along the upper reaches of the river and around the lake from which it flows.

In later years—especially after the discovery of the Montana gold-mines had drawn to the upper valleys of the Missouri an adventurous, gold-seeking population, who scoured the mountains in all directions—rumors of burning plains, spouting springs, great lakes and other natural wonders, came down from the unknown regions up the Yellowstone. And not content with these, the imagination was freely drawn on, and the treasure valleys of the Arabian Nights were rivalled, if not reproduced. Our over-venturous party, hotly pursued by Indians, escaped, report said, by travelling night after night by the brilliant light of a huge diamond providentially exposed on a mountain. A lost trapper turned up after protracted wandering in this mysterious region, his pockets stuffed with nuggets of gold gathered in a stream which he could never find again. More astounding still was a valley which instantly petrified whatever entered it. Rabbits and sage-hens, even Indians were standing about there, like statuary, among thickets of petrified sage-brush, whose stony branches bore diamonds, rubies, sapphires, emeralds and other gems by the thousand, as large as walnuts. “I tell you, sir,” said one who had been there, to Colonel Raynolds, “it is true, for I gathered a quart myself and sent them down the country.”

The first earnest attempt to explore the valley of the upper Yellowstone was made in 1859, by Colonel Raynolds, of the Corps of Engineers. His expedition passed entirely around the Yellowstone basin, but could not penetrate it. In his report to the War Department, he says:

“It was my original desire to go from the head of Wind River to the head of the Yellowstone, keeping on the Atlantic slope, thence down the Yellowstone, passing the lake, and across by the Gallatin to the three forks of the Missouri. Bridger said at the outset that this would be impossible, and that it would be necessary to cross over to the headwaters of the Columbia and back again to the Yellowstone. I had not previously believed that crossing the main crest twice would be more easily accomplished than the transit over what was in effect only a spur; but the view from our first camp settled the question adversely to my opinion at once. Directly across our route lies a basaltic ridge, rising not less than 5,000 feet above us, its walls apparently vertical, with no visible pass or even cañon. On the opposite side of this are the headwaters of the Yellowstone. Bridger remarked triumphantly and forcibly on reaching this spot, 'I told you you could not go through. A bird can't fly over that without taking a supply of grub along.' I had no reply to offer, and mentally conceded the accuracy of the information of 'the old man of the mountains.'

* * * * *

"After this obstacle had thus forced us over on the western slope of the Rocky Mountains, an effort was made to recross and reach the district in question, but although it was June, the immense body of snow baffled all our exertions, and we were compelled to content ourselves with listening to marvellous tales of burning plains, immense lakes, and boiling springs, without being able to verify these wonders. I know of but two white men who claim to ever have visited this part of the Yellowstone Valley—James Bridger and Robert Meldrum. The narratives of both these men are very remarkable, and Bridger, in one of his recitals, described an immense boiling spring, that is a perfect counterpart of the Geysers of Iceland. As he is uneducated, and had probably never heard of the existence of such natural marvels elsewhere, I have little doubt that he spoke of that which he had actually seen. The burning plains described by these men may be volcanic, or, more probably, burning beds of lignite similar to those on Powder River, which are known to be in a state of ignition.... Had our attempt to enter this district been made a month later in the season, the snow would have mainly disappeared, and there would have been no insurmountable obstacles to overcome.

"I cannot doubt, therefore, that at no very distant day the mysteries of this region will be fully revealed, and though small in extent, I regard the valley of the upper Yellowstone as the most interesting unexplored district of our widely expanded country."

Ten years after Colonel Raynolds's unsuccessful attempt to solve the problem of the Yellowstone, a small party under Messrs. Cook and Folsom ascended the river to the lake, and crossed over the divide into the Geyser Basin of the Madison. No report, we believe, was published of their discoveries. At any rate, the general public were indebted for their first knowledge of the marvels of this region to an expedition organized in the summer of 1870 by some of the officials and leading citizens of Montana.

This company, led by General Washburn, the Surveyor-General of the Territory, and accompanied by a small escort of United States cavalry under Lieutenant G. C. Doane, left Fort Ellis toward the latter part of August, and entered the valley of Yellowstone River on the 23d. During the next thirty days they explored the cañons of the Yellowstone and the shores of Yellowstone Lake; then crossing the mountains to the headwaters of the Madison, they visited the geyser region of Firehole River, and ascended that stream to its junction with the Madison, along whose valley they returned to civilization, confident, as their historian wrote, that they had seen "the greatest wonders on the Continent," and "convinced that there was not on the globe another region where, within the same limits, nature had crowded so much of grandeur and majesty, with so much of novelty and wonder."

Mr. Langford's account of this expedition, published in the second volume of Scribner's Monthly, and the report of Lieutenant Doane, printed some time after by the United States Government, (Ex. Doc. No. 51, 41st Congress,) gave to the world the first authentic information of the marvels of this wonderful region. Though their route lay through a terrible wilderness, and most of the party were but amateur explorers at best, only one (Mr. Everts) met with a serious mishap. This gentleman's story of his separation from the company, and his thirty-seven days of suffering and perilous wandering, is one of the most thrilling chapters of adventure ever written.

The path fairly broken, and the romance of the Yellowstone shown to have a substantial basis in reality, it was not long before others were ready to explore more fully the magnificent scenery and the strange and peculiar phenomena described by the adventurers of 1870. As soon as the following season was sufficiently advanced to admit of explorations among the mountains, the Chief Engineer of the Military Department of the Missouri, Brevet Colonel John W. Barlow, set out for a two months' survey of the Yellowstone Basin, under special orders from General Sheridan. His route lay along the river to the lake; thence along the northern shore of the lake to the hot springs on its western bank; thence across the mountains westward to the Geyser Basins of Firehole River, which he ascended to its source in Madison Lake; thence to Heart Lake, the source of Snake River; thence across the mountains to Bridger's Lake, in the valley of the Upper Yellowstone. Descending this stream to where it enters Yellowstone Lake, he returned by the east shore of the lake to Pelican Creek; thence across the country to the Falls of the Yellowstone; thence over the mountains to the East Fork of the Yellowstone, which he followed to its junction with the main stream.

In the meantime, a large and thoroughly-organized scientific party, under Dr. F. V. Hayden, U. S. geologist, were making a systematic survey of the region traversed by Colonel Barlow. The work done by this party is briefly summarized by Dr. Hayden as follows:

"From Fort Ellis, we passed eastward over the divide, between the drainage of the Missouri and Yellowstone, to Bottlers' Ranch. Here we established a permanent camp, leaving all our wagons and a portion of the party. A careful system of meteorological observations was kept at this locality for six weeks. From Bottlers' Ranch we proceeded up the valley of the Yellowstone, surveyed the remarkable hot springs on Gardiner's River, The Grand Cañon, Tower Falls, Upper and Lower Falls of the Yellowstone, thence into the basin proper, prepared charts of all the Hot Spring groups, which were very numerous, and continued up the river to the lake. We then commenced a systematic survey of the lake and its surroundings. Mr. Schönborn, with his assistant, made a careful survey of the lake and the mountains from the shore, and Messrs. Elliott and Carrington surveyed and sketched its shore-lines from the water in a boat. Careful soundings were also made, and the greatest depth was found to be three hundred feet. From the lake I proceeded, with Messrs. Schönborn, Peale, and Elliott to the Firehole Valley, by way of East Fork of the Madison; then ascended the Firehole Valley. We made careful charts of the Lower and Upper Geyser Basin, locating all the principal springs, and determining their temperatures. We then returned over the mountains by way of the head of Firehole River, explored Madison Lake, Heart Lake, etc. After having completed our survey of the lake, we crossed over to the headwaters of the East Fork by way of the valley of Pelican Creek, explored the East Fork to its junction with the main Yellowstone, and thence to Bottlers' Ranch, which we reached on the 28th of August. From this place we passed down the Yellowstone, through the lower cañon, to the mouth of Shield's River, to connect our work with that of Colonel Wm. F. Raynolds, in 1860. From there we returned to Fort Ellis."

It is safe to say that no exploring expedition on this continent ever had a more interesting field of investigation, or ever studied so many grand, curious and wonderful aspects of nature in so short a time.

Chapter 3

FORT ELLIS TO BOTTLERS' RANCH

The Yellowstone tourist leaves the confines of civilization at Fort Ellis. This frontier military post, situated near the head of the beautiful and fertile valley of the East Gallatin, commands the valleys of the Yellowstone and the three forks of the Missouri—the finest and most productive portion of Montana. On the east and north are ranges of hills and mountains which form the divide between the waters of the Yellowstone and the Missouri. On the south and west, the beautiful Valley of the Gallatin. Abundant vegetation, beautiful scenery, streams of pure water flowing down the mountain-sides and across the plains on every hand, and a climate that can hardly be surpassed in any country, combine to make this pleasant station one of the most charming places on the continent.

For the first six miles the road from Fort Ellis to the wonder-land of the Yellowstone Valley follows the general course of the East Gallatin, up steep acclivities and through the defiles of a hilly country to the crest of the divide. The road here takes advantage of a natural pass between hills that rise from six hundred to twelve hundred feet above the road, itself considerably more elevated than the summit of the White Mountains. From the tops of the hills on either side the view is wonderfully fine in every direction. To the west lies the Gallatin Valley, with its cordon of snow-capped peaks, its finely-timbered water courses, and its long, grassy declivities, dotted with the habitations of pioneers, and blooming with the fruits of industry. To the eastward lies the beautiful Valley of the Yellowstone, not yet laid under tribute to man. On the further side of this valley—the bed of an ancient lake—the eye takes in at a glance one of the most symmetrical and remarkable ranges of mountains in all the West. Indeed, Dr. Hayden says, in describing them:

"Several of my party who had visited Europe regarded this range as in no way inferior in beauty to any in that far-famed country. A series of cone-shaped peaks, looking like gigantic pyramids, are grouped along the east side of the valley for thirty or forty miles, with their bald, dark summits covered with perpetual snow, the vegetation growing thinner and smaller as we ascend the almost vertical sides, until, long before reaching the summits, it has entirely disappeared. On all sides deep gorges have been gashed out by aqueous forces cutting through the very core of the mountains, and forming those wonderful gulches which only the hardy and daring miner has ventured to explore. This range, which is called on the maps Snowy Mountains, forms the great water-shed between two portions of the Yellowstone River, above and below the first cañon, and gives origin to some of the most important branches of that river. From the summit of Emigrant Peak, one of the highest of these volcanic cones, one great mass of these basaltic peaks can be seen as far as the eye can reach, rising to the height of 10,000 to 11,000 feet above the sea. Emigrant Peak, the base of which is cut by the Yellowstone River, is 10,629 feet above tide-water, while the valley plain near Bottlers' Ranch, on the opposite side of the river, was found to be 5,925 feet. This splendid group of peaks rises 5,000 feet and upward above the valley of the Yellowstone."

About three miles from the divide the road strikes the valley of Trail Creek, a small-sized trout-stream of great clearness and purity, flowing southeastward to the Yellowstone, between high hills wooded at the summits. Approaching the river, the country becomes more and more volcanic in appearance, masses of basaltic lava cropping out from the high ridges on the right and left. Many of these masses show a perpendicular front of several hundred feet, with projections resembling towers, castles and the like. Several miles away on the right, is Pyramid Mountain, a snow-capped peak. Farther to the south is a long range of mountains, also covered with snow, even in midsummer.

On the left of the valley the foot hills bear abundant verdure, the highest summits being covered with a vigorous growth of pines. Trail Creek enters the Yellowstone about thirty miles from Fort Ellis. Ten miles further up the Yellowstone is Bottlers' Ranch, the last abode of civilized man in this direction.

The Bottler brothers, who have established themselves here, belong to that numerous class of pioneers who are satisfied only when their field of operation is a little in advance of civilization, exposed to privation and danger, yet possessing advantages for hunting, trapping and fishing not enjoyed by men content to dwell in safety. These, however are not their only occupations. They have under cultivation large fields of wheat, potatoes and other crops, possess extensive herds of cattle, and make large quantities of butter, for which they find a ready market in the mining camps of Emigrant Gulch across the river, which at this point is a very rapid stream, about three hundred feet wide and four feet deep on the riffles at low water.

Of this part of the valley Dr. Hayden says: "It is about fifteen miles long, and will average three miles in width; it is well watered, soil fertile, and in every respect one of the most desirable portions of Montana. We may not look for any districts favorable for agriculture in the Yellowstone Valley above the second cañon; but this entire lake basin seems admirably adapted for grazing and for the cultivation of the usual crops of the country. The cereals and the roots have already been produced in abundance, especially whcat and potatoes. The mountains on either side are covered with snow, to a greater or less extent, all the year, which in melting feeds the numerous little streams that flow down the mountain-sides in the Yellowstone. Hundreds of springs flow out of the terraces. One terrace near Bottlers' Ranch gives origin to fifty springs within a mile, and then, all aggregating together in the river bottom, form a large stream. Thus there is the greatest abundance of water for irrigation, or for any of the purposes of settlement. The elevation of the valley at this ranch is 4,925 feet, and this may be regarded as the average in altitude. But a small portion of it is occupied as yet, but the time is not far distant when the valley will be covered with fine farms and the hills with stock. It will always be a region of interest, from the fact that it is probably the upper limit of agricultural effort in the Yellowstone Valley."

Chapter 4

BOTTLERS' RANCH TO GARDINER'S RIVER

At Bottlers' Ranch the wagon road terminates. For the first ten miles beyond, the trail runs along the west bank of the river through the wildest imaginable scenery of rock, river and mountain. The path is narrow, rocky and uneven, frequently leading over steep hills of considerable height. From the top of one of these, a bold mountain spur coming down to the water's edge, the view up the valley is very fine, embracing the river fringed with cottonwoods, the foot hills covered with luxuriant, many-tinted herbage, and over all the snow-crowned summits of the distant mountains. Above this point the valley opens out to a "bottom" of large extent and great beauty. Across the river the steep lava mountains come close to the stream, their lofty fronts covered with stunted timber. A large portion of the bottom land is subject to overflow by the numerous mountain streams that come in from the right, and bears an abundance of grass, in many places waist high. The river is skirted with shrubbery and cedars, the latter having thick trunks, too short for ordinary lumber, yet of beautiful grain for small cabinet work, and susceptible of exquisite finish.

At the head of this valley is the second cañon of the Yellowstone, granite walls rising on either side to the height of a thousand feet or more, and the river dashing through the narrow gorge with great velocity. Seen from the lofty mountain spur over which the trail is forced to pass, the bright green color of the water, and the numerous ripples, capped with white foam, as the roaring torrent rushes around and over the multitude of rocks that have fallen from above into the channel, present a most picturesque appearance. Above the cañon, which is about a mile in length, the valley widens slightly, then narrows so as to compel the traveller to cross a ridge, on whose summit lies a beautiful lake. Descending to the valley again the road traverses a tract of level bottom land, a mile or two wide, covered with a heavy growth of sage-brush. Throughout all this portion of its course, the Yellowstone is abundantly stocked with trout of the largest variety known this side the Rocky Mountains.

Some ten miles above the second cañon on the edge of the river valley is Cinnabar Mountain, whose weather-beaten side presents one of the most singular freaks of nature in the world. Two parallel vertical walls of rock, fifty feet wide, traverse the mountain from base to summit, and project to the height of three hundred feet for a distance of fifteen hundred feet. The sides are as even as if wrought by line and plumb. The rock between the walls and on either side has been completely worn away. Speaking of this curious formation, Mr. Langford says:

"We had seen many of the capricious works wrought by erosion upon the friable rocks of Montana, but never before upon so majestic a scale. Here an entire mountain-side, by wind and water, had been removed, leaving as the evidences of their protracted toil these vertical projections, which, but for their immensity, might as readily be mistaken for works of art as of nature. Their smooth sides, uniform width and height, and great length, considered in connection with the causes which had wrought their insulation, excited our wonder and admiration. They were all the more curious because of their dissimilarity to any other striking objects in natural scenery that we had ever seen or heard of. In future years, when the wonders of the Yellowstone are incorporated into the family of fashionable resorts, there will be few of its attractions surpassing in interest this marvellous freak of the elements."

According to the observations of Dr. Hayden, the mountain is formed of alternate beds of sandstone, limestone, and quartzites, elevated to a nearly vertical position by those internal forces which acted in ages past to lift the mountain ranges to their present heights. Standing at the base and looking up the sides of the mountain, the geologist could not but be filled with wonder at the convulsions which threw such immense masses of rocks into their present position. Ridge after ridge extends down the steep sides of the mountain like lofty walls, the intervening softer portions having been washed away, leaving the harder layers projecting far above. In one place the rocks incline in every possible direction, and are crushed together in the utmost confusion. Between the walls at one point is a band of bright brick-red clay, which has been mistaken for cinnabar, and hence the name Cinnabar Mountain. The most conspicuous ridge is composed of basalt, which must have been poured out on the surface when all the rocks were in a horizontal position. For reasons best known to himself, one of the first explorers of this region gave these parallel ridges the title of "Devil's Slide."

"The suggestion was unfortunate," writes the historian of the Expedition, "as, with more reason perhaps, but with no better taste, we frequently had occasion to appropriate other portions of the person of his Satanic Majesty, or of his dominion, in signification of the varied marvels we met with. Some little excuse may be found for this in the fact that the old mountaineers and trappers who preceded us had been peculiarly lavish in the use of the infernal vocabulary. Every river and glen and mountain had suggested to their imaginations some fancied resemblance to portions of a region which their pious grandmothers had warned them to avoid. It is common for them, when speaking of this region, to designate portions of its physical features, as "Firehole Prairie,"—the "Devil's Den,"—"Hell Roaring River," etc.—and these names, from a remarkable fitness of things, are not likely to be speedily superseded by others less impressive."

These "impressive" titles stand in curious contrast with the fanciful names bestowed in this region by Capts. Lewis and Clarke,—Wisdom River, Philosophy River, Philanthropy Creek, and the like.

From the Devil's Slide to the mouth of Gardiner's River, twelve miles, the ground rises rapidly, passing from a dead level alkali plain, to a succession of plateaus covered slightly with a sterile soil. Evidences of volcanic action begin to be frequent: old craters converted into small lakes appear here and there, prettily fringed with vegetation, and covered with waterfowl. Scattered over the hills and through the valleys are numerous beautiful specimens of chalcedony and chips of obsidian. Many of the chalcedonies are geodes, in which are crystals of quartz; others contain opal in the centre and agate on the exterior; and still others have on the outside attached crystals of calcite.

Chapter 5

HOT SPRINGS OF GARDINER'S RIVER

Ten miles above the Devil's Slide, Gardiner's River, a mountain torrent twenty yards wide, cuts through a deep and gloomy gorge and enters the Yellowstone at the lower end of the Third Cañon.

At this point the Yellowstone shrinks to half its usual size, losing itself among huge granite boulders, which choke up the stream and create alternate pools and rapids, crowded with trout. Worn into fantastic forms by the washing water, these immense rock masses give an aspect of peculiar wildness to the scenery. But the crowning wonder of this region is the group of hot springs on the slope of a mountain, four miles up the valley of Gardiner's River. The first expedition passed on without seeing them, but they could not escape the vigilance of the scientific company that followed.

The lower reaches of the valley of Gardiner's River, and the enclosing hillsides, are strewn with volcanic rock, having the appearance of furnace cinder. The tops of the rounded hills are covered with fragments of basalt and conglomerate, whose great variety of sombre colors add much to the appearance of desolation which characterizes the valley. Here and there are stagnant lakes fifty to a hundred yards in diameter, apparently occupying ancient volcanic vents. Crossing a barren, elevated region two miles in extent, and three or four hundred feet above the river-bed, the trail descends abruptly to a low "bottom" covered with a thick calcareous crust, deposited from hot springs, now for the most part dry. At one point, however, a large stream of hot water, six feet wide and two feet deep, flows swiftly from beneath the crust, its exposed portion clearly revealed by rising steam. The quantity of water flowing from this spring is greater than from any other in this region; its temperature ranges from 126° to 132° Fah. A little further above are three or four other springs near the margin of the river. These have nearly circular basins, six to ten feet in diameter, and a temperature not above 120°. Already these springs have become the resort of invalids, who speak highly of the virtues of the waters. A short distance up the hill are abundant remains of springs, which in time past must have been very active. For nearly a mile the steep hill-side is covered with a thick crust of spring deposits, which, though much decomposed and overgrown with pines and cedars, still bear traces of the wonderful forms displayed in the vicinity of the active springs further up the hill. Ascending the hill, Dr. Hayden's party came suddenly and unexpectedly upon these marvellous deposits, which they agreed in pronouncing one of the finest displays of natural architecture in the world. The snowy whiteness of the deposit, which has the form of a frozen cascade, at once suggested the name of White Mountain Hot Spring. The springs now in active operation cover an area of about one square mile, while the rest of the territory, three or four square miles in extent, is occupied by the remains of springs which have ceased to flow. Small streams flow down the sides of the Snowy Mountain in channels lined with oxide of iron of the most delicate tints of red. Others show exquisite shades of yellow, from a deep, bright sulphur, to a dainty cream-color. Still others are stained with shades of green, all these colors as brilliant as the brightest aniline dyes. The water after rising from the spring basin flows down the sides of the declivity, step by step, from one reservoir to another, at each one of them losing a portion of its heat, until it becomes as cool as spring-water. Within five hundred feet of its source Dr. Hayden's party camped for two days by the side of the little stream formed by the aggregated waters of these hot springs, and found the water most excellent for drinking as well as for cooking purposes. It was perfectly clear and tasteless, and harmless in its effects. During their stay here all the members of the party, as well as the soldiers comprising their escort, enjoyed the luxury of bathing in these most elegantly carved natural bathing-pools; and it was easy to select, from the hundreds of reservoirs, water of any desired temperature. These natural basins vary somewhat in size, but many of them are about four by six feet in diameter, and one to four feet in depth. Their margins are beautifully scalloped, and adorned with a natural beadwork of exquisite beauty.

The level or terrace upon which the principal active springs are located, is about midway up the sides of the mountain, covered with the sediment. Still farther up are the ruins of what must have been at some period more active springs than any at present known. The sides of the mountain for two or three hundred feet in height, are thickly encrusted with calcareous deposit, originally ornamented with elegant sculpturing, like the bathing pools below; but atmospheric agencies, which act readily on the lime, have obliterated all their delicate beauty.

Bathing-Pools (Diana's Bath.)

The largest living spring is near the outer margin of the main terrace. Its dimensions are twenty-five feet by forty, and its water so perfectly transparent that one can look down into the beautiful ultramarine depth to the very bottom of the basin. Its sides are ornamented with coral-like forms of a great variety of shades, from pure white to a bright cream yellow, while the blue sky reflected in the transparent water gives an azure tint to the whole which surpasses all art. From various portions of the rim, water flows out in moderate quantities over the sides of the hill. Whenever it gathers into a channel and flows quite swiftly, basins with sides from two to eight feet high are formed with their ornamental designs proportionately coarse; but when the water flows slowly, myriads of little basins are formed, one below another, with a semblance of irregular system. The water holds in solution a great amount of lime, with some soda, alumina and magnesia, which are slowly deposited as the water flows down the sides of the mountain. Underneath the sides of many of the pools are rows of exquisitely-ornamented stalactites, formed by the dripping of the water over the margins. All these springs have one or more centres of ebullition which is constant, though seldom rising more than four or five inches above the surface. The ebullition is due mainly to the emission of carbonic acid gas. The springs in the centre of the main basin are probably all at the boiling point—194° at this elevation. Being inaccessible, however, it is impossible to determine their actual temperature. The hottest that could be reached was 162° Fah. The terrace immediately above the main basin is bordered by a long rounded ridge with a fissure extending its whole length, its interior lined with beautiful crystals of pure sulphur. Only hot vapors and steam issue from this fissure, though the bubbling and gurgling of water far beneath the surface can be distinctly heard. Back of this ridge are several small springs which throw up geyser-like jets of water intermittently to the height of three feet.

On the west side of this deposit, about one-third of the way up the White Mountain from the river and terrace, where was once the theatre of many active springs, old chimneys or craters are scattered thickly over the surface, and there are several large holes and fissures leading to vast caverns below. The crust gives off a dull hollow sound beneath the tread, and the surface gives indistinct evidence of having been adorned with the beautiful pools or basins already described. At the base of the principal terrace is a large area covered with shallow pools some of them containing water, with all the ornamentations perfect, while others are fast going to decay, and the decomposed sediment is as white as snow. On this sub-terrace is a remarkable cone about 50 feet in height and 20 feet in diameter at the base. Its form has suggested the name of Liberty Cap. It is undoubtedly the remains of an extinct geyser. The water was forced up with considerable power, and probably without intermission, building up its own crater until the pressure beneath was exhausted, and the spring gradually closed itself over at the summit and perished. No water flows from it at the present time. The layers of lime were deposited around the cap like the layers of straw on a thatched roof, or hay on a conical stack. Not far from the Liberty Cap is a smaller cone, called, from its form, the "Bee-hive." These springs are constantly changing their position; some die out, others burst out in new places. On the northwest margin of the main terrace are examples of what have been called oblong mounds. There are several of them in this region, extending in different directions, from fifty to one hundred and fifty yards in length, from six to ten feet high, and from ten to fifteen feet broad at the base.

The Liberty Cap.

There is in all cases a fissure from one end of the summit to the other, usually from six to ten inches wide, from which steam sometimes issues in considerable quantities, and on walking along the top one can hear the water seething and boiling below like a cauldron. The inner portion of the shell, as far down as can be seen, is lined with a hard, white enamel-like porcelain; in some places beautiful crystals of sulphur have been precipitated from the steam. These mounds have been built up by a kind of oblong fissure-spring in the same way that the cones have been constructed. The water, continually spouting up, deposited sediment around the edges of the fissure until the force was exhausted, and then the calcareous basin was rounded up something like a thatched roof by overlapping layers. Near the upper terrace, which is really an old rim, are a number of these extinct, oblong geysers, some of which have been broken down so as to show them to be mere shells or caverns, now the abode of wild animals. Dr. Hayden attempted to enter one of them, and found it full of sticks and bones which had been carried in by wild beasts; and swarms of bats flitted to and fro. Some of the mounds have been worn away so that sections are exposed, showing the great number and thickness of the overlapping layers of sediment. Many mounds are overgrown with pine-trees, which must be at least eighty or a hundred years old. Indeed, the upper part of this mountain appears like a magnificent ruin of a once flourishing village of these unique structures, now fast decomposing, yet beautiful and instructive in their decay. One may now study the layers of deposit, sometimes thousands on a single mound, as he would the rings of growth in a tree. How long a period is required to form one of these mounds, or to build up its beautiful structure, there is no data for determining. On the middle terrace, where the principal portion of the active springs are, some of the pine-trees are buried in sediment apparently to the depth of six or eight feet. All of them are dead at the present time. There is, however, evidence enough around the springs to show that the mineral-water is precipitated with great rapidity. It is probable that all the deposits in the immediate vicinity of the active springs are constantly changing from the margin of the river to the top of the White Mountain and return. The deposits upon the summit are extensive, though now there is very little water issuing from the springs there, and that is of low temperature. Quantities of steam are ever ascending from the springs, and on damp mornings the entire slope of the mountain is enveloped in clouds of vapor.

"But," observes Dr. Hayden, in summing up his account of this indescribable locality, "it is to the wonderful variety of exquisitely delicate colors that this picture owes the main part of its attractiveness. The little orifices from which the hot water issues are beautifully enamelled with the porcelain-like lining, and around the edges a layer of sulphur is precipitated. As the water flows along the valley, it lays down in its course a pavement more beautiful and elaborate in its adornment than art has ever yet conceived. The sulphur and the iron, with the green microscopic vegetation, tint the whole with an illumination of which no decoration-painter has ever dreamed. From the sides of the oblong mound, which is here from 30 to 50 feet high, the water has oozed out at different points, forming small groups of the semicircular, step-like basins.

"Again, if we look at the principal group of springs from the high mound above the middle terrace, we can see the same variety of brilliant coloring. The wonderful transparency of the water surpasses anything of the kind I have ever seen in any other portion of the world. The sky, with the smallest cloud that flits across it, is reflected in its clear depths, and the ultramarine colors, more vivid than the sea, are greatly heightened by the constant gentle vibrations. One can look down into the clear depths and see, with perfect distinctness, the minutest ornament on the inner sides of the basins; and the exquisite beauty of the coloring and the variety of forms baffle any attempt to portray them, either with pen or pencil. And then, too, around the borders of these springs, especially those of rather low temperature, and on the sides and bottoms of the numerous little channels of the streams that flow from these springs, there is a striking variety of the most vivid colors. I can only compare them to our most brilliant aniline dyes—various shades of red, from the brightest scarlet to a bright rose tint; also yellow, from deep-bright sulphur, through all the shades, to light cream-color. There are also various shades of green, from the peculiar vegetation. These springs are also filled with minute vegetable forms, which under the microscope prove to be diatoms, among which Dr. Billings discovers Palmella and Oscillara. There are also in the little streams that flow from the boiling springs great quantities of a fibrous, silky substance, apparently vegetable, which vibrates at the slightest movement of the water, and has the appearance of the finest quality of cashmere wool. When the waters are still these silken masses become incrusted with lime, the delicate vegetable threads disappear, and a fibrous, spongy mass remains, like delicate snow-white coral."

The antiquity of these springs is a question of great interest, yet difficult of solution. When were these immense deposits begun? On the margin of the mountain, high above the present position of the hot springs, is a bed of white, or yellowish white limestone, from fifty to a hundred and fifty feet thick. It is regularly stratified and the jointing is complete. There is a belt a mile long and one fourth of a mile wide, covered with cubical masses of this rock that have fallen down the slope of the mountain. These immense blocks, fifty to one hundred feet in each dimension, appear as if the mass had slowly fallen down as the underlying rocks were worn away. So thickly is this belt covered with these huge masses that it is with the greatest difficulty one can walk across it. It would seem that this bed must at one time have extended over a portion or all of the valley of Gardiner's River. Much of the rock is very compact, and would make beautiful building-stone, on account of its close texture and color, and it could be converted into the whitest of lime. If the rocks are examined, however, over a considerable area, they are found to possess all the varieties of structure of a hot-spring deposit. Some portions are quite spongy, and decompose readily; others are made up of very thin laminæ, regular or wavy; enough to show the origin of the deposit without a doubt. But in what manner was it formed? Dr. Hayden believes that the limestone was precipitated in the bottom of a lake, which was filled with hot-springs, much as the calcareous matter is laid down in the bottom of the ocean at the present time. Indeed, portions of the rock do not differ materially from the recent limestones now forming in the vicinity of the West India Islands. The deposit was evidently laid down on a nearly level surface, with a moderately uniform thickness, and the strata are horizontal. Since this group of strata was formed, the country has been elevated, and the valley of Gardiner's River has been carved out, so that the commencement of the period of activity of these springs must date back to a period merging on, but just prior to, the present geological period—probably at the time of the greatest action of the volcanic forces in this region.

Classed with reference to their chemical constituents, the springs here and elsewhere in the Yellowstone Valley are of two kinds: those in which lime predominates, and those in which silica is most abundant. The springs of Gardiner's River are mainly the former. Where does the lime come from? The geology of the country surrounding the springs shows already that there is underneath the spring deposits, at least a thickness of 1,500 feet, of carboniferous limestone; and if the origin of the heat which so elevates the temperature of the waters of these springs is as deep seated as is generally supposed, the heated waters have ample time and space for dissolving the calcareous rocks through which they flow.

Chapter 6

GARDINER'S RIVER TO GRAND CAÑON

About a mile above the springs, Gardiner's River separates into three branches—the East, Middle, and West Forks, which rise high up in the mountains, among perpetual snows. They wind their way across a broad plateau covered mostly with a dense growth of pines, but with some broad, open, meadow-like spots, which, seen from some high mountain peak, lend a rare charm to the landscape. After gathering a sufficient supply of water, they commence wearing their channels down into the volcanic rocks, deeper and deeper as they descend. Each one has its water-fall, which would fill an artist with enthusiasm. From the high ridge between the East and Middle Forks a fine view is obtained of the surrounding country.

Far to the southwest are lofty peaks covered with snow, rising to the height of 10,000 feet, and forming a part of the magnificent range of mountains that separates the Yellowstone from the sources of the Gallatin. From this high ridge one can look down into the chasm of the Middle Fork, carved out of the basalt and basaltic conglomerates to the depth of 500 to 800 feet, with nearly vertical sides. In the sides of this cañon, as well as those of the East Fork, splendid examples of basaltic columns are displayed, as perfect as those of the celebrated Fingal's Cave. They usually appear in regular rows, vertical, five and six sided, but far more sharply cut than elsewhere seen in the West, though occasionally the columns are spread out in the form of a fan. Sometimes there are several rows, usually about fifty feet high, one above the other, with conglomerate between.

The cañon is about 500 yards from margin to margin at top, but narrows down until on the bottom it is not more than forty yards wide. At one point the water pours over a declivity of 300 feet or more, forming a most beautiful cascade. The direct fall is over 100 feet. The constant roar of the water is like that of a train of cars in motion. The pines are very dense, usually of moderate size, and among them are many open spaces, covered with stout grass, sometimes with large sage-bushes. Upon the high hills the vegetation is remarkably luxuriant, indicating great fertility of soil, which is usually very thick, and made up mostly of degraded igneous rocks. Above the falls the rows of vertical, basaltic columns continue in the walls of the cañon, and they may well be ranked among the remarkable wonders of this rare wonderland. The lower portion of the cañon is composed of rather coarse igneous rocks, which have a jointage and a style of weathering like granite. The West Fork rolls over a bed of basalt, which is divided into blocks that give the walls the appearance of mason-work on a gigantic scale. Below the falls the river has cut the sides of the mountain, exposing a vertical section 400 feet high, with the same irregular jointage.

South of the hot springs is a round dome-like mountain, rising 2,100 feet above them, or 8,500 feet above the sea. Its summit commands a prospect from thirty to fifty miles in every direction. To the north and west stands a group of lofty peaks over 10,000 feet above the sea, and covered with huge masses of snow. These peaks form a part of the range that separates the waters of the Gallatin from those of the Yellowstone. Farther on to the southward are the peaks of the head of the Madison, and in the interval one black mass of pine forest, covering high plateaus, with no point rising over 8,500 feet above the sea—the whole region being more or less wavy or rolling, interspersed here and there with beautiful lakes a few hundred yards in diameter; and here and there a bright-green grassy valley through which little streams wind their way to the large rivers. In one of these lakes the explorers saw the greatest abundance of yellow water-lily, which blooms in great profusion on the surface of all the mountain lakes of the Yellowstone

Basin. On the east side of Gardiner's Cañon, and west of the Yellowstone, is a sort of wave-like series of ridges, with broad, open, grassy interspaces, with many groves of pines. These ridges gradually slope down to the Yellowstone, northeast. Far to the east and north is one jagged mass of volcanic peaks, some of them snow-clad, others bald and desolate to the eye. Far to the south, dimly outlined on the horizon, may be seen the three Tetons and Madison Peak—monarchs of all the region. A grander view could not well be conceived.

Leaving Gardiner's River, Dr. Hayden's party ascended the broad slope of the dividing ridge between that river and the streamlets which flow into the Yellowstone. Immense boulders of massive granite, considerably rounded, are a marked feature of the country about the entrance of the East Fork. One of these, a mass of red feldspathic granite, is twenty-five feet thick and fifty feet long. The high wavy ridge, 9,000 feet above the sea, is composed of beds of steel-gray and brown sandstone and calcareous clay, in which are numerous impressions of deciduous leaves. Vast quantities of silicified wood of great perfection and beauty are scattered all over the surface. In some cases long trees have been turned to agate, the rings of growth as perfectly shown as in recent wood. The soil is very thick, and covered with luxuriant vegetation.

"We were travelling through this region in the latter part of the month of July," writes Dr. Hayden, "and all the vegetation seemed to be in the height of its growth and beauty. The meadows were covered densely with grass and flowers of many varieties, and among the pines were charming groves of poplars, contrasting strongly by their peculiar enlivening foliage with the sombre hue of the pines. The climate was perfect, and in the midst of some of the most remarkable scenery in the world, every hour of our march only increased our enthusiasm.

"The climate during the months of June, July, and August, in this valley, cannot be surpassed in the world for its health-giving powers. The finest of mountain water, fish in the greatest abundance, with a good supply of game of all kinds, fully satisfy the wants of the traveller, and render this valley one of the most attractive places of resort for invalids or pleasure-seekers in America."

From the summit of the ridge the party descended to the valley of the Yellowstone, nearly opposite the mouth of the East Fork of that river. The road was a rough one. During the period of volcanic action in this region, the sedimentary rocks were crumpled into high, sharp, wave-like series of ridges; from innumerable fissures, igneous matter was poured out over the surface cooling into basalt; and from volcanic vents was also thrown out, into the great lake, rock fragments and volcanic dust, which were arranged by the water and cemented into a breccia. Deep into these ridges the little streams have cut their channels, forming what should be called valleys, rather than cañons, with almost vertical sides. These ravines, 500 to 800 feet deep, covered mostly with grass or trees, occur in great numbers, many of them entirely dry at present, but attesting the presence and power, at no very remote period, of aqueous forces compared with which those of the present are utterly insignificant.

Before studying this portion of the Yellowstone Valley, it may be well to retrace our steps to the mouth of Gardiner's River, to explore the Third Cañon of the Yellowstone, so far as possible, and the rest of its interesting valley up to this point.

As already noticed, the country about the mouth of Gardiner's River is desolate and gloomy. The hill-slopes are covered with sage brush, the constant sign of arid soil, and grass is scarce. This is the first poor camping-place on the route. The cañon being impassable, the trail passes to the right, crossing several high mountain-spurs, over which the way is much obstructed by fallen timber, and reaching at last a high rolling plateau. This elevated tract is about thirty miles in extent, with a general declivity to the north. Its surface is an undulating prairie, dotted with groves of pine and aspen. Numerous lakes are scattered throughout its whole extent, and great numbers of springs, which flow down the slopes, are lost in the volume of the Yellowstone. The river breaks through this plateau in a winding cañon over 2,000 feet in depth—the middle cañon of the Yellowstone rolling over volcanic boulders in some places, and in others forming still pools of seemingly fathomless depth. At one point it dashes to and fro, lashed to a white foam on its rocky bed; at another, where a deep basin occurs in the channel, it subsides into a crystal mirror. Numerous small cascades are seen tumbling from the rocky walls at different points and the river appears from the lofty summits a mere ribbon of foam in the immeasurable distance below. Standing on the brink of the chasm the heavy roaring of the imprisoned river comes to the ear only in a sort of hollow, hungry growl, scarcely audible from the depths. Lofty pines on the bank of the stream "dwindle to shrubs in dizziness of distance." Everything beneath, says Lieut. Doane, has a weird and deceptive appearance. The water does not look like water, but like oil. Numerous fish-hawks are seen busily plying their vocation, sailing high above the waters, and yet a thousand feet below the spectator. In the clefts of the rocks down, hundreds of feet down, bald eagles have their eyries, from which one can see them swooping still farther into the depths to rob the ospreys of their hard-earned trout. It is grand, gloomy, and terrible; a solitude peopled with fantastic ideas; an empire of shadows and of turmoil.

The plateau formation is of lava, generally in horizontal layers, as it cooled in a surface flow, yet upheaved in many places into wave-like undulations. Occasionally granite shafts protrude through the strata, forming landmarks of picturesque form. Like dark icebergs stranded in an ocean of green, they rise high above the tops of the trees in wooded districts, or stand out grim and solid on the grassy expanse of the prairies.

Near the head of the Third Cañon a stream flows into the Yellowstone from the northeast, bearing the sonorous title, Hell-Roaring River. It is quite a large stream, rising high among the mountains, and flowing with tremendous impetuosity down the deep gorges. The mountains on either side come close down to the channel of the Yellowstone, and are among the most rugged in this rugged region. A huge peak of this sort, composed of stratified gneiss, with deep strata of massive red and grey granite, stands at the mouth of Hell-Roaring River, and takes to itself the same imposing name. A short distance above the mouth of Hell Roaring River, the East Fork of the Yellowstone comes in from the southeast. Its sources are high up among the most rugged and inaccessible portions of the basaltic range, several jagged peaks which rise from 10,000 to 11,000 feet above the sea.

Extinct Geyser, East Fork of the Yellowstone.

"The summits of these high peaks," observes Mr. Hayden, "are all close, compact trachyte, while all around the sides are built up walls of stratified conglomerate. It is plain that all of them are the nuclei of old volcanoes. The trachyte may sometimes be concealed by the conglomerates, but I am inclined to think that each one has formed a centre of effusion. Large quantities of silicified wood are found among the conglomerates, mostly inclosed in the volcanic cement, evidently thrown out of the active craters with the fragments of basalt. My impression is, that when the old volcanoes disgorged their contents into the great lake of waters around, they threw out also portions from the sedimentary formations, and thus the silicified wood comes from the Tertiary or Cretaceous beds, which may have formed the upper part of the walls of the crater. At any rate, these woods belong to the Coal Series of the West, and they are scattered profusely among the conglomerates. Interlaced among the massive beds of volcanic conglomerates are some layers of a light-grey or whitish sandy clay, which show that the whole breccia or conglomerates, with the intercalated layers of clay or sand, were deposited in water like any sedimentary water rocks."

Interesting ruins of ancient springs abound in this valley. Mr. Hayden describes one, a very curious mammiform mound of calcareous deposit, about forty feet high, built up by overlapping layers like those of Liberty Cap on Gardiner's River.

"This cone is a complete ruin. No water issues from it at the present time, and none of the springs in the vicinity are above the ordinary temperature of brook-water; sulphur, alum, and other chemical deposits are abundant. This old ruin is a fine example of the tendency of the cone to close up its summit in its dying stages. The top of the cone is somewhat broken; but it is eighteen feet in diameter at this time, and near the centre there is a hole or chimney two inches in diameter, plainly a steam-vent. This marks the closing history of this spring. The inner portions of this small chimney are lined with white enamel, thickly coated with sulphur, which gives it a sulphur-yellow hue. The base upon which the cone rests varies in thickness. On the east side huge masses have been broken off, exposing a vertical wall twenty feet high, built up of thin horizontal laminæ of limestone. On the west side the wall is not quite as high, perhaps eight or ten feet. It would seem, therefore, that it was at first an overflowing spring, depositing thin horizontal layers until it built up a broad base ten to twenty feet in height; then it gradually became a spouting spring, building up with overlapping layers like the thatch on a house, until it closed itself at the top and ceased."

In the tongue that runs down between the junction of the East Fork and the Yellowstone, there is a singular butte cut off from the main range, which at once attracts the traveller's attention. The basis or lower portion of the butte is granite, while the summit is capped with the modern basalt, and the débris on the sides and at the base is remarkable in quantity, and has very much the appearance of an anthracite coal-heap. This butte will always form a conspicuous landmark, not only on account of its position, but also from its peculiar shape and structure.

Just below the junction of the East Fork the first and only bridge across the Yellowstone was constructed in 1870 for the accommodation of miners bound for the "diggings" on Clark's Fork. It was a work of considerable boldness, as the river is some two hundred feet wide, and flows with great rapidity over its narrow and rocky channel.

A short distance above the bridge, on the west side of the Yellowstone, is a splendid exhibition of black micaceous gneiss, forming a vertical wall on the right side of a little creek, while on the left the entire mass of the hills for miles in extent is composed of the usual igneous rocks. Through these rocks the stream, now not more than four feet wide and six inches deep, has cut a channel from two hundred to four hundred yards wide, through the hardest rocks to a depth varying from five hundred to a thousand feet!

Further up the Yellowstone, on the same side, are a number of wonderful ravines and cañons carved in like manner into the very heart of the mountains. Most conspicuous of these is the Cañon of Tower Greek. Before reaching that stream, however, Column Rock, a noticeable feature in a landscape of great extent and beauty, demands at least a passing notice. Column Cliff would be a more appropriate name, since it extends along the east bank of the river upwards of two miles. Says Mr. Langford, whose observations were made from the west side: "At the distance from which we saw it, we could compare it in appearance to nothing but a section of the Giant's Causeway. It was composed of successive pillars of basalt overlying and underlying a thick stratum of cement and gravel resembling pudding-stone. In both rows, the pillars, standing in close proximity, were each about thirty feet high and from three to five feet in diameter. This interesting object, more from the novelty of its formation and its beautiful surroundings of mountain and river scenery than anything grand or impressive in its appearance, excited our attention, until the gathering shades of evening reminded us of the necessity of selecting a suitable camp."

Tower Creek rises in the high divide between the valleys of the Missouri and the Yellowstone, and flows for about ten miles through a cañon so deep and gloomy that it has earned the appellation, "Devil's Den." About two hundred yards above its entrance into the Yellowstone, the stream pours over an abrupt descent of one hundred and fifty-six feet, forming one of the most beautiful falls to be found in any country. These falls are about 260 feet above the level of the Yellowstone at the junction, and are surrounded with columns of volcanic breccia, rising fifty feet above the falls and extending down to the foot, standing like gloomy sentinels, or like gigantic pillars at the entrance of some grand temple. "One could almost imagine," says Dr. Hayden, "that the idea of the Gothic style of architecture had been caught from such carvings of nature."

Speaking of the symmetry of some of these columns, Mr. Langford says:

"Some resemble towers, others the spires of churches, and others still shoot up as lithe and slender as the minarets of a mosque. Some of the loftiest of these formations, standing like sentinels upon the very brink of the fall, are accessible to an expert and adventurous climber. The position attained on one of their narrow summits, amid the uproar of waters and at a height of 250 feet above the boiling chasm, as the writer can affirm, requires a steady head and strong nerves; yet the view which rewards the temerity of the exploit is full of compensations. Below the fall the stream descends in numerous rapids, with frightful velocity, through a gloomy gorge, to its union with the Yellowstone. Its bed is filled with enormous boulders, against which the rushing waters break with great fury.

Many of the capricious formations wrought from the shale excite merriment as well as wonder. Of this kind especially was a huge mass sixty feet in height, which, from its supposed resemblance to the proverbial foot of his Satanic Majesty, we called the "Devil's Hoof." The scenery of mountain, rock, and forest surrounding the falls is very beautiful. Here, too, the hunter and fisherman can indulge their tastes with the certainty of ample reward. As a half-way resort to the greater wonders still farther up the marvellous river, the visitor of future years will find no more delightful resting-place. The name of "Tower Falls," which we gave it, was suggested by some of the most conspicuous features of the scenery."

The sides of the chasm are worn into caverns lined with variously-tinted mosses, nourished by clouds of spray which rise from the cataract; while above, and to the left, a spur from the great plateau rises over all, with a perpendicular front of 400 feet. The fall is accessible both at the brink and at the foot, and fine views can be obtained from either side of the cañon. In appearance it strongly resembles Minnehaha, but is several times as high, and the volume of water is at least eight times as great.

The Devil's Hoof.

In the basin a large petrified log was found imbedded in the débris. "Nothing," says Lieutenant Doane, "can be more chastely beautiful than this lovely cascade, hidden away in the dim light of overshadowing rocks and woods, its very voice hushed to a low murmur, unheard at the distance of a few hundred yards. Thousands might pass by within a half mile and not dream of its existence; but once seen, it passes to the list of most pleasant memories." Along the Yellowstone, near the mouth of Tower Creek, is a system of small mineral springs distributed for a distance of two miles in the bottom of the deep cañon through which the river runs. Several of these springs have a temperature at the boiling point; many are highly sulphurous, holding, in fact, more sulphur than they can carry in solution, and depositing it in yellowish beds along their courses. Several of them are impregnated with iron, alum, and other substances. Their sulphurous fumes can be detected at the distance of half a mile. The excess of sulphur in the rock-walls of the cañon give a brilliant yellow color to the rocks in many places. The formation is usually very friable, falling with a natural slope to the edge of the stream, but occasionally masses of a more solid nature project from the wall in curious shapes of towers, minarets, and the like; while over all the solid ledge of trap, with its dark and well-defined columns, makes a rich and beautiful border inclosing the pictured rocks above.

This is the mouth of the Grand Canon of the Yellowstone.

Chapter 7

Over Mount Washburn to the Falls of the Yellowstone

The Upper or Grand Cañon of the Yellowstone extends from the mouth of Tower Creek to the foot of the Great Fall, a distance of twenty miles. It is impassable throughout its entire length, and accessible to the water's edge only at few points and by dint of severe labor. The trail ascends the divide between Tower Creek and the Yellowstone, skirting for six or eight miles the cañon of Tower Creek. The ground rises rapidly and is much broken by creek-beds running parallel with the river. Following the highest ridges, the first explorers reached at last a point whence they could overlook the Grand Cañon cleaving the slopes and breaking through the lofty mountain ranges in front. Here they caught their first glimpse of a phenomenon afterwards to become a familiar sight to them. Through the mountain gap formed by the cañon, and on the interior slopes some twenty miles distant, an object appeared which drew a simultaneous expression of wonder from every one in the party. It was a column of steam, rising from the dense woods to the height of several hundred feet. They had all heard fabulous stories of this region, and were somewhat skeptical of appearances. At first it was pronounced a fire in the woods, but presently some one noticed that the vapor rose in regular puffs, as if expelled with a great force. Then conviction was forced upon them. It was, indeed, a great column of steam, puffing away on the lofty mountain side, with a roaring sound, audible at a long distance, even through the heavy forest. A hearty cheer rang out at this discovery, and they pressed onward with renewed enthusiasm.

The highest peak of this ridge was named by the first company who climbed it—Mount Washburn—in honor of their leader. The view from its summit is "grand beyond description;" yet some conception of its grandeur can be formed, let us hope, from the graphic review of its more striking features by Lieutenant Doane.

"Looking northward, the great plateau stretches away from the base of the mountain to the front and left with its innumerable groves and sparkling waters, a variegated landscape of surpassing beauty, bounded on its extreme verge by the cañons of the Yellowstone. The pure atmosphere of this lofty region causes every outline of tree, rock, or lakelet to be visible with wonderful distinctness, and objects twenty miles away appear as if very near at hand. Still further to the left the snowy ranges on the headwaters of Gardiner's River stretch away to the westward, joining those on the head of the Gallatin, and forming, with the Elephant's Back, a continuous chain, bending constantly to the south, the rim of the Yellowstone Basin. On the verge of the horizon appear, like mole-hills in the distance, and far below, the white summits above the Gallatin Valley. These never thaw during the summer months, though several thousand feet lower than where we now stand upon the bare granite, and no snow visible near, save in the depths of shaded ravines. Beyond the plateau to the right front is the deep valley of the East Fork bearing away eastward, and still beyond, ragged volcanic peaks, heaped in inextricable confusion, as far as the limit of vision extends. On the east, close beneath our feet, yawns the immense gulf of the Grand Cañon, cutting away the bases of two mountains in forcing a passage through the range. Its yellow walls divide the landscape nearly in a straight line to the junction of Warm Spring (Tower) Creek below. The ragged edges of the chasm are from two hundred to five hundred yards apart, its depth so profound that the river bed is nowhere visible. No sound reaches the ear from the bottom of the abyss; the sun's rays are reflected on the farther wall and then lost in the darkness below. The mind struggles and then falls back upon itself, despairing in the efforts to grasp by a single thought the idea of its immensity. Beyond, a gentle declivity, sloping from the summit of the broken range, extends to the limit of vision, a wilderness of unbroken pine forest.

"Turning southward, a new and strange scene bursts upon the view. Filling the whole field of vision, and with its boundaries in the verge of the horizon, lies the great volcanic basin of the Yellowstone—nearly circular in form, from fifty to seventy-five miles in diameter; and with a general depression of about 2,000 feet below the summits of the great ranges which form its outer rim. Mount Washburn lies in the point of the circumference, northeast from the centre of the basin; far away in the southwest, the three great Tetons on Snake River fill another space in the circle; connecting these two highest are crescent ranges, one westward and south, past Gardiner's River and the Gallatin, bounding the lower Madison, thence to the Jefferson and by the Snake River range to the Tetons; another eastward and south, a continuous range by the head of Rose Bud, inclosing the sources of the Snake, and joining the Tetons beyond. Between the south and west points, this vast circle is broken through in many places for the passage of the rivers; but a single glance at the interior slopes of the ranges shows that a former complete connection existed, and that the great basin has been one vast crater of a now extinct volcano. The nature of the rocks, the steepness and outline of the interior walls, together with other peculiarities to be mentioned hereafter, render this conclusion a certainty. The lowest point in this great amphitheatre lies directly in front of us, and about eight miles distant: a grassy valley, branching between low ridges, running from the river toward the centre of the basin. A small stream rises in this valley, breaking through the ridges to the west in a deep cañon, and falling into the channel of the Yellowstone, which here bears in a northeast course, flowing in view as far as the confluence of the small stream, thence plunged into the Grand Cañon, and hidden from sight. No falls can be seen, but their location is readily detected by the sudden disappearance of the river; beyond this open valley the basin appears to be filled with a succession of low, converging ridges, heavily timbered, and all of about an equal altitude.

"To the south appears a broad sheet of water—the Yellowstone Lake. Across the Grand Cañon, on the slope of the great mountain wall, is the steam jet seen this morning; and in the next ravine beyond it are six more of inferior volume. Still farther south are others, to the number of perhaps twenty, and to the southwest more of them, scattered over the vast expanse of the basin, rising from behind the wooded hills in every direction. The view in this respect strongly resembles that from the Alleghanies, where they overlook iron and coal districts, with all their furnaces in active operation, save that one looks in vain here for the thrifty towns, country villas, steamboats, and railroad depots."

Does this picture seem overdrawn? The briefer and less enthusiastic description of Dr. Hayden confirms its truth, though he does not accept in full Lieutenant Doane's interpretation of it. He says, in his official report:

"The view from the summit of Mount Washburn is one of the finest I have ever seen, and although the atmosphere was somewhat obscured by smoke, yet an area of fifty to one hundred miles radius in every direction could be seen more or less distinctly. We caught the first glimpse of the great basin of the Yellowstone, with the lake, which reminded one much, from its bays, indentations, and surrounding mountains, of Great Salt Lake. To the south are the Tetons, rising high above all the rest, the monarchs of all they survey, with their summits covered with perpetual snow. To the southwest an immense area of dense pine forests extends for one hundred miles without a peak rising above the black, level mass. A little farther to the southwest and west are the Madison Mountains, a lofty, grand, snow-capped range, extending far to the northward. Nearer and in full view, to the west commence the bold peaks of the Gallatin Range, extending northward as far as the eye can reach. To the north we get a full view of the valley of the Yellowstone, with the lofty ranges that wall it in. Emigrant Peak, and the splendid group of mountains of which it is a part, can be clearly seen, and lose none of their marvellous beauty of outline, view them from what point we may. To the north and east the eye scans the most remarkable chaotic mass of peaks of the most rugged character, apparently without system, yet sending their jagged summits high up among the clouds. Farther distant are somewhat more regular ranges, snow-covered, probably the Big Horn. But with all this magnificent scenery around us from every side, the greatest beauty was the lake, in full view to the southeast, set like a gem amid the high mountains, which are literally bristling with peaks, many of them capped with snow. These are all of volcanic origin, and the fantastic shapes which many of them have assumed under the hand of time, called forth a variety of names from my party. There were two of them that represented the human profile so well that we called them the "Giant's Face "and "Old Man of the Mountain." These formed good landmarks for the topographer, for they were visible from every point of the basin."

As regards the geological character of the country seen from Mount Washburn, Dr. Hayden observes, in discussing the geology of the region:

"We may say, in brief, that the entire basin of the Yellowstone is volcanic. I am not prepared to pronounce it a crater, with a lake occupying the inner portion, while the mountains that surround the basin are the ruins of this great crater; but, at a period not very remote in the geological past, this whole country was a scene of wonderful volcanic activity. I regard the hot springs so abundant all over the valley as the last stages of this grand scene. Hot springs, geysers, etc., are so intimately connected with what we usually term volcanoes that their origin and action admit of the same explanation. Both undoubtedly form safety-valves or vents for the escape of the powerful forces that have been generated in the interior of the earth since the commencement of our present period; the true volcanic action has ceased, but the safety-valves are the thousands of hot springs all over this great area. I believe that the time of the greatest volcanic activity occurred during the Pliocene period—smoke, ashes, fragments of rock, and lava poured forth from thousands of orifices into the surrounding waters. Hundreds of cones were built up, fragments of which still remain; and around them were arranged by the water the dust and fragments of rock, the *ejectamenta* of these volcanoes, in the form of the conglomerate or breccia as we find it now. These orifices may have been of every possible form—rounded or oblong, mere fissures, perhaps, extending for miles, and building up their own crater rims as the hot springs build up their rounded, conical peaks or oblong mounds at the present time."

Leaving Mount Washburn, with its summit piles of basalt, and its precipitous slope scattered with agates and beautiful fragments of sardonyx, chalcedony, and malachite, let us descend to the valley.

The trail pursues a tortuous way to avoid the fallen timber and the dense groves of pine, descending the almost vertical inner sides of the rim of the Yellowstone Basin, to the valley of a small creek. Two or three miles down this stream is a hideous glen, filled with sulphurous vapor emitted from six or eight boiling springs of great size and activity. Mr. Langford says of this unsavory place:

"It looked like nothing earthly we had ever seen, and the pungent fumes which filled the atmosphere were not unaccompanied by a disagreeable sense of possible suffocation. Entering the basin cautiously, we found the entire surface of the earth covered with the incrusted sinter thrown from the springs. Jets of hot vapor were expelled through a hundred natural orifices with which it was pierced, and through every fracture made by passing over it. The springs themselves were as diabolical in appearance as the witches' caldron in Macbeth, and needed but the presence of Hecate and her weird band to realize that horrible creation of poetic fancy. They were all in a state of violent ebullition, throwing their liquid contents to the height of three or four feet. The largest had a basin twenty by forty feet in diameter. Its greenish-yellow water was covered with bubbles, which were constantly rising, bursting, and emitting sulphurous gas from various parts of its surface. The central spring seethed and bubbled like a boiling caldron. Fearful volumes of vapor were constantly escaping it. Near it was another, not so large, but more infernal in appearance. Its contents, of the consistency of paint, were in constant, noisy ebullition. A stick thrust into it, on being withdrawn, was coated with lead-colored slime a quarter of an inch in thickness. Nothing flows from this spring. Seemingly, it is boiling down. A fourth spring, which exhibited the same physical features, was partly covered by an overhanging ledge of rock. We tried to fathom it, but the bottom was beyond the reach of the longest pole we could find. Rocks cast into it increased the agitation of its waters. There were several other springs in the group, smaller in size, but presenting the same characteristics.

"The approach to them was unsafe, the incrustation surrounding them bending in many places beneath our weight,—and from the fractures thus created would ooze a sulphury slime of the consistency of mucilage. It was with great difficulty that we obtained specimens from the natural apertures with which the crust is filled,—a feat which was accomplished by one only of our party, who extended himself at full length upon that portion of the incrustation which yielded the least, but which was not sufficiently strong to bear his weight while in an upright position, and at imminent risk of sinking into the infernal mixture, rolled over and over to the edge of the opening, and with the crust slowly bending and sinking beneath him, hurriedly secured the coveted prize."

Getting a Specimen.

"There was something so revolting in the general appearance of the springs and their surroundings—the foulness of the vapors, the infernal contents, the treacherous incrustation, the noisy ebullition, the general appearance of desolation, and the seclusion and wildness of the location—that, though awestruck, we were not unreluctant to continue our journey without making them a second visit."

Once more our amateur explorers had recourse to their western vocabulary, and bestowed on this unhappy locality the title, "Hell-broth Springs"—which, says the historian of the expedition, "fully expressed our appreciation of their character."

The following season this remarkable group of springs was thoroughly examined by the party under Dr. Hayden. That careful observer says:

"They are evidently diminishing in power, but the rims all around reveal the most powerful manifestations far back in the past. Sulphur, copper, alum, and soda cover the surface. There is also precipitated around the borders of some of the mud springs a white efflorescence, probably nitrate of potash. These springs are located on the side of the mountain, nearly 1,000 feet above the margin of the cañon, but extend along into the level portions below.... One of these springs was bubbling quite briskly, but had a temperature of only 100°. Near it is a turbid spring of 170°. In the valley are a large number of turbid, mud, and boiling springs, with temperatures from 175° to 185°. There are a number of springs that issue from the side of the mountain, and the waters, gathering into one channel, flow into the Yellowstone, The number of frying or simmering springs is great. The ground in many places, for several yards in every direction, is perforated like a sieve, and the water bubbles by with a simmering noise. There is one huge boiling spring which deposits fine black mud all around the sides. The depth of the crater of this spring, its dark, gloomy appearance, and the tremendous force which it manifested in its operations, led us to name it the "Devil's Caldron." There are a large number of springs here, but no true geysers. It is plainly the last stages of what was once a most remarkable group. Extending across the cañon on the opposite side of the Yellowstone, interrupted here and there, this group of springs extends for several miles, forming one of the largest deposits of silica, but only here and there are there signs of life. Many of the dead springs are mere basins, with a thick deposit of iron on the sides, lining the channel of the water that flows from them. These vary in temperature from 98° to 120°. The highest temperature was 192°. The steam-vents are very numerous, and the chimneys are lined with sulphur. Where the crust can be removed, we find the underside lined with the most delicate crystals of sulphur, which disappear like frost-work at the touch.

Still there is a considerable amount of solid amorphous sulphur. The sulphur and the iron, with the vegetable matter which is always very abundant about the springs, give, through the almost infinite variety of shades, a most pleasing and striking picture. One of the mud springs, with a basin twenty by twenty-five feet, and six feet deep, is covered with large bubbles or puffs constantly bursting with a thud. There are a number of high hills in this vicinity entirely composed of the hot-spring deposits, at least nine-tenths silica, appearing snowy-white in the distance; one of the walls is 175 feet high, and another about 70 feet. They are now covered to a greater or less extent with pines. Steam is constantly issuing from vents around the base and from the sides of these hills. There is one lake 100 by 300 yards, with a number of bubbling and boiling springs rising to the surface. Near the shore is one of the sieve-springs, with a great number of small perforations, from which the water bubbles up with a simmering noise; temperature, 188°. This group really forms one of the great ruins."

A short day's march from Hell-broth Spring brings the traveller to a little stream flowing into the Yellowstone, between the upper and the lower fall. From its rapid and tumultuous flow, the first explorers called it Cascade Creek. Just before its union with the Yellowstone it traverses a gloomy gorge cut through a kind of volcanic sandstone, largely made up of fragments of obsidian and other igneous rocks cemented with volcanic ash. This rock is worn by the water into so many fantastic shapes and cavernous recesses, that—with their usual poverty of invention and tartarean taste—the first observers straightway gave the uncanny channel over to the Prince of Darkness, and dubbed it the Devil's Den. A mile below this gorge the stream flows over a series of ledges, making a cascade as beautiful as its previous course has been weird and ugly. There is first a fall of five feet, which is immediately succeeded by another of fifteen, into a pool as clear as amber, nestled beneath overarching rocks. Here the stream lingers as if half reluctant to continue its course, then gracefully emerges from the grotto, and, veiling the rocks down an abrupt descent of eighty-four feet, passes rapidly on to the Yellowstone.

The Devil's Den.

For a wonder, this charming fall has received a corresponding name—Crystal Cascade. An infinite variety of volcanic specimens, quartz, feldspar, mica, granites, lavas, basalts, composite crystals—in fact, everything, from asbestos to obsidian, is represented by fragments in the bed of this stream. At the foot of the gorge and on the margin of the Yellowstone stands a high promontory of concretionary lava, literally filled with volcanic butternuts. Many of these are loose, and can be taken out of the rock with the hand;

broken open, they are invariably hollow, and lined with minute quartz crystals of various tints. This rare formation occurs frequently in the great basin.

Chapter 8

The Grand Cañon and the Falls

No language," says Dr. Hayden, "can do justice to the wonderful grandeur and beauty of the Grand Cañon." It has no parallel in the world. Through the eye alone can any just idea be gained of its strange, awful, fascinating, unearthly blending of the majestic and the beautiful; and, even in its visible presence, the mind fails to comprehend the weird and unfamiliar, almost incredible scenes it reveals. Says Mr. Langford: "The brain reels as we gaze into this profound and solemn solitude. We shrink from the dizzy verge appalled, glad to feel the solid earth under our feet, and venture no more, except with forms extended, and faces barely protruding over the edge of the precipice. The stillness is horrible. Down, down, down, we see the river attenuated to a thread, tossing its miniature waves, and dashing, with puny strength, against the massive walls which imprison it. All access to its margin is denied, and the dark gray rocks hold it in dismal shadow. Even the voice of its waters in their convulsive agony cannot be heard. Uncheered by plant or shrub, obstructed with massive boulders and by jutting points, it rushes madly on its solitary course. The solemn grandeur of the scene surpasses description. The sense of danger with which it impresses you is harrowing in the extreme. You feel the absence of sound, the oppression of absolute silence. If you could only hear that gurgling river, if you could see a living tree in the depth beneath you, if a bird would fly past, if the wind would move any object in the awful chasm, to break for a moment the solemn silence that reigns there, it would relieve that tension of the nerves which the scene has excited, and you would rise from your prostrate condition and thank God that he had permitted you to gaze, unharmed, upon this majestic display of natural architecture. As it is, sympathizing in spirit with the deep gloom of the scene, you crawl from the dreadful verge, scared lest the firm rock give way beneath and precipitate you into the horrid gulf."

"The fearful descent into this terrific cañon," Mr. Langford adds, "was accomplished with great difficulty by Messrs. Hauser and Stickney, at a point about two miles below the falls. By trigonometrical measurement they found the chasm at that point to be 1,190 feet deep. Their ascent from it was perilous, and it was only by making good use of hands and feet, and keeping the nerves braced to the utmost tension, that they were enabled to clamber up the precipitous rocks to a safe landing-place."

Lieutenant Doane also made the descent, somewhat further down the river, accompanied by one of his company. Selecting the channel of a small creek, they scrambled down its steep descent, wading in the stream.

"On entering the ravine, we came at once to hot springs of sulphur, sulphate of copper, alum, steam jets, etc., in endless variety, some of them of very peculiar form. One of them in particular, of sulphur, had built up a tall spire, standing out from the slope of the wall like an enormous horn, with hot water trickling down its sides. The creek ran on a bed of solid rock, in many places smooth and slippery, in others obstructed by masses of débris from the overhanging cliffs of the sulphureted limestone above. After descending for three miles in the channel we came to a sort of bench or terrace, the same one seen previously in following down the creek from our first camp in the basin. Here we found a large flock of mountain sheep, very tame, and greatly astonished, no doubt, at our sudden appearance. McConnell killed one and wounded another, whereupon the rest disappeared, clambering up the steep walls with a celerity truly astonishing. We were now 1,500 feet below the brink. From here the creek channel was more precipitous, and for a mile we made our way down over masses of rock and fallen trees, splashing in warm water, ducking under cascades, and skirting close against sidelong places to keep from falling into boiling caldrons in the channel. After four hours of hard labor we reached the bottom of the gulf and the margin of the Yellowstone, famished with thirst, wet and exhausted. The river-water here is quite warm, and of a villainously alum and sulphurous taste. Its margin is lined with all kinds of chemical springs, some depositing craters of calcareous rock, others muddy, black, blue, slaty, or reddish water. The internal heat renders the atmosphere oppressive, though a strong breeze draws through the cañon. A frying sound comes constantly to the ear, mingled with the rush of the current. The place abounds with sickening and purgatorial smells. We had come down the ravine at least four miles, and looking upward the fearful wall appeared to reach the sky. It was about three o'clock P.M., and stars could be distinctly seen, so much of the sunlight was cut off from entering the chasm. Tall pines on the extreme verge appeared the height of two or three feet. The cañon, as before said, was in two benches, with a plateau on either side, about half way down. This plateau, about a hundred yards in width, looked from below like a mere shelf against the wall; the total depth was not less than 2,500 feet, and more probably 3,000. There are perhaps other cañons longer and deeper than this one, but surely none combining grandeur and immensity with such peculiarity of formation and profusion of volcanic or chemical phenomena."

The history of this tremendous chasm is not hard to read. Ages ago this whole region was the basin of an immense lake. Then it became a centre of volcanic activity; vast quantities of lava was erupted, which, cooling under water, took the form of basalt; volumes of volcanic ash and rock-fragments were thrown out from the craters from time to time, forming breccia as it sunk through the water and mingled with the deposits from silicious springs. Over this were spread the later deposits from the waters of the old lake. In time the country was slowly elevated, and the lake was drained away. The easily eroded breccia along the river channel was cut out deeper and deeper as the ages passed, while springs and creeks and the falling rain combined to carve the sides of the cañon into the fantastic forms they now present, by wearing away the softer rock, and leaving the hard basalt and the firmer hot-spring deposits standing in massive columns and Gothic pinnacles. The basis material of the old hot spring deposits in silica, originally white as snow, are now stained by mineral waters with every shade of red and yellow—from scarlet to rose color, from bright sulphur to the daintiest tint of cream. When the light falls favorably on these blended tints the Grand Cañon presents a more enchanting and bewildering variety of forms and colors than human artist ever conceived.

The erosion was practically arrested at the upper end of the cañon by a sudden transition from the softer breccia to hard basalt, and the falls are the result. From below the Upper Fall the vertical wall of basalt can be clearly seen passing diagonally across the rim. The Lower Fall was formed in the same way.

"A grander scene than the lower cataract of the Yellowstone," writes Mr. Langford, "was never witnessed by mortal eyes. The volume seemed to be adapted to all the harmonies of the surrounding scenery. Had it been greater or smaller it would have been less impressive. The river, from a width of two hundred feet above the fall, is compressed by converging rocks to one hundred and fifty feet, where it takes the plunge. The shelf over which it falls is as level and even as a work of art. The height, by actual line measurement, is a few inches more than 350 feet. It is a sheer, compact, solid, perpendicular sheet, faultless in all the elements of grandeur and picturesque beauties. The cañon which commences at the upper fall, half a mile above this cataract, is here a thousand feet in depth. Its vertical sides rise grey and dark above the fall to shelving summits, from which one can look down into the boiling, spray-filled chasm, enlivened with rainbows, and glittering like a shower of diamonds. From a shelf protruding over the stream, 500 feet below the top of the cañon, and 180 above the verge of the cataract, a member of our company, lying prone upon the rock, let down a cord, with a stone attached, into the gulf, and measured its profoundest depths. The life and sound of the cataract, with its sparkling spray and fleecy foam, contrasts strangely with the sombre stillness of the cañon a mile below. There all was darkness, gloom, and shadow; here all was vivacity, gayety, and delight. One was the most unsocial, the other the most social scene in nature. We could talk, and sing, and whoop, waking the echoes with our mirth and laughter in presence of the falls, but we could not thus profane the silence of the cañon. Seen through the cañon below the falls, the river for a mile or more is broken by rapids and cascades of great variety and beauty.

"Between the Lower and Upper Falls the cañon is two hundred to nearly four hundred feet deep. The river runs over a level bed of rock, and is undisturbed by rapids until near the verge of the lower fall. The upper fall is entirely unlike the other, but in its peculiar character equally interesting. For some distance above it the river breaks into frightful rapids. The stream is narrowed between the rocks as it approaches the brink, and bounds with impatient struggles for release, leaping through the stony jaws, in a sheet of snow-white foam, over a precipice nearly perpendicular, 115 feet high.[1] Midway in its descent the entire volume of water is carried, by the sloping surface of an intervening ledge, twelve or fifteen feet beyond the vertical base of the precipice, gaining therefrom a novel and interesting feature. The churning of the water upon the rocks reduces it to a mass of foam and spray, through which all the colors of the solar spectrum are reproduced in astonishing profusion. What this cataract lacks in sublimity is more than compensated by picturesqueness. The rocks which overshadow it do not veil it from the open light. It is up amid the pine foliage which crowns the adjacent hills, the grand feature of a landscape unrivalled for beauties of vegetation as well as of rock and glen. The two confronting rocks, overhanging the verge at the height of a hundred feet or more, could be readily united by a bridge, from which some of the grandest views of natural scenery in the world could be obtained—while just in front of, and within reaching distance of the arrowy water, from a table one-third of the way below the brink of the fall, all its nearest beauties and terrors may be caught at a glance."

"We rambled around the falls and cañon two days, and left them with the unpleasant conviction that the greatest wonder of our journey had been seen."

A few scattered sentences, culled from Dr. Hayden's calmly scientific account of the falls, will suffice to show that Mr. Langford's description "o'ersteps not the modesty of nature."

---

[1] Mr. Langford appears to have underestimated this fall. The report of the U. S. Geological Survey gives the height as 140 feet.

Upper Falls of the Yellowstone.

"Above the Upper Falls the Yellowstone flows through a grassy, meadow-like valley, with a calm, steady current, giving no warning, until very near the falls, that it is about to rush over a precipice 140 feet, and then, within a quarter of a mile, again to leap down a distance of 350 feet.

"From any point of view the Upper Falls are extremely picturesque and striking. The entire volume of water seems to be, as it were, hurled off of the precipice with the force which it has accumulated in the rapids above, so that the mass is detached into the most beautiful snow-white, bead-like drops, and as it strikes the rocky basin below, it shoots through the water with a sort of ricochet for the distance of 200 feet. The whole presents in the distance the appearance of a mass of snow-white foam. On the sides of the basalt walls there is a thick growth of vegetation, nourished by the spray above, which extends up as far as the moisture can reach.... After the waters roll over the upper descent, they flow with great rapidity over the apparently flat rocky bottom, which spreads out to nearly double its width above the falls, and continues thus until near the Lower Falls, when the channel again contracts, and the waters seem, as it were, to gather themselves into one compact mass and plunge over the descent of 350 feet in detached drops of foam as white as snow; some of the large globules of water shoot down like the contents of an exploded rocket.... The entire mass of the water falls into a circular basin, which has been worn into the hard rock, so that the rebound is one of the magnificent features of the scene.... It is a sight far more beautiful, though not so grand or impressive as that of Niagara Falls. A heavy mist always rises from the water at the foot of the falls, so dense that one cannot approach within 200 or 300 feet, and even then the clothes will be drenched in a few moments. Upon the yellow, nearly vertical wall of the west side, the mist mostly falls, and for 300 feet from the bottom the wall is covered with a thick matting of mosses, sedges, grasses, and other vegetation of the most vivid green, which have sent their small roots into the softened rocks, and are nourished by the ever-ascending spray. At the base and quite high up on the sides of the cañon, are great quantities of talus, and through the fragments of rocks and decomposed spring deposits may be seen the horizontal strata of breccia."

On his return down the opposite or eastern side of the river, Colonel Barlow descended to the foot of the Lower Fall for the purpose of exploring the cañon. He says: "I expected this to be an undertaking of great difficulty and attended with some danger, but entering a sharp and narrow gorge or fissure in the side of the cañon, immediately below the great fall, I found the descent much easier than was anticipated. It proved to be very steep, but the rock being solid, with projecting angles, there was little danger to a careful climber. A slope of loose and finely broken rock, a hundred feet in height, moist from the falling spray, terminated the descent. Sliding to the bottom of this slope, I stood at the foot of the great fall, 350 feet below its crest, the walls of the cañon rising 700 feet. My first impression on beholding this fall from below was one of disappointment; it did not appear as high as I expected. The fall, however, was grand, and presented a symmetrical and unbroken sheet of white foam, set in dark masses of rock, while rainbows were formed in the spray from almost every point of view. The steep rocks near the falls, constantly wet with rising mist, were covered with vegetation of an intensely green color. The river below runs with the velocity of a torrent, rushing down declivities, spinning round sharp angles, and dashing itself into spray at every turn. I found it impossible to follow the bed of the stream, the steep and slippery side affording no footing whatever, and crumbling at the slightest touch."

Chapter 9

FROM THE FALLS TO THE LAKE

Half a mile above the Upper Fall the Yellowstone gives no intimation of its approaching career of wildness and grandeur. It rolls peacefully between low verdant banks and over pebbly reaches or spaces of quicksand, with beautiful curves and a majestic motion. Its waters are clear and cold, and of the emerald hue characteristic of Niagara. Great numbers of small springs, fed by the slowly melting snows of the mountains, flow from the densely wooded foot-hills, irrigating the "bottoms," and sustaining a growth of grass and flowers that clothes the lowlands with freshness and vividness of color. Everything terrific, diabolic, volcanic, would seem to have been left behind. The first hint to the contrary is given by a pretty little rivulet, a yard wide and a few inches deep, clear as crystal, winding along through the rank grass to join the Yellowstone. It looks like any clear-watered mountain stream; but a single taste shows that it has a different origin. It is strongly charged with alum—hence its name, Alum Creek—and its source is in a remarkable group of sulphur and alum springs two or three miles further on,—that is, about ten miles above the falls.

All about these springs are evidences of volcanic action in great variety and profusion. Mr. Langford says:

"The region was filled with boiling springs and craters. Two hills, each 300 feet high, and from a quarter to half a mile across, had been formed wholly of the sinter thrown from adjacent springs—lava, sulphur, and reddish-brown clay. Hot streams of vapor were pouring from crevices scattered over them. Their surfaces answered in hollow intonations to every footstep, and in several places yielded to the weight of our horses. Steaming vapor rushed hissingly from the fractures, and all around the natural vents large quantities of sulphur in crystallized form, perfectly pure, had been deposited. This could be readily gathered with pick and shovel. A great many exhausted craters dotted the hill-side. One near the summit, still alive, changed its hues like steel under the process of tempering, to every kiss of the passing breeze. The hottest vapors were active beneath the incrusted surface everywhere. A thick leathern glove was no protection to the hand exposed to them. Around these immense thermal deposits, the country, for a great distance in all directions, is filled with boiling springs, all exhibiting separate characteristics.

"The most conspicuous of the cluster is a sulphur spring twelve by twenty feet in diameter, encircled by a beautifully scolloped sedimentary border, in which the water is thrown to a height of from three to seven feet. The regular formation of this border, and the perfect shading of the scollops forming it, are among the most delicate and wonderful freaks of nature's handiwork. They look like an elaborate work of art. This spring is located at the western base of Crater Hill, above described, and the gentle slope around it for a distance of 300 feet is covered to considerable depth with a mixture of sulphur and brown lava. The moistened bed of a small channel, leading from the spring down the slope, indicated that it had recently overflowed.

"A few rods north of this spring, at the base of the hill, is a cavern whose mouth is about seven feet in diameter, from which a dense jet of sulphurous vapor explodes with a regular report like a high-pressure engine. A little farther along we came upon another boiling spring, seventy feet long by forty wide, the water of which is dark and muddy and in unceasing agitation.

"About a hundred yards distant we discovered a boiling alum spring, surrounded with beautiful crystals, from the border of which we gathered a quantity of alum, nearly pure, but slightly impregnated with iron. The violent ebullition of the water had undermined the surrounding surface in many places, and for the distance of several feet from the margin had so thoroughly saturated the incrustation with its liquid contents, that it was unsafe to approach the edge. As one of our company was unconcernedly passing near the brink, the incrustation suddenly sloughed off beneath his feet. A shout of alarm from his comrades aroused him to a sense of his peril, and he only avoided being plunged into the boiling mixture by falling suddenly backward at full length upon the firm portion of the crust, and rolling over to a place of safety. His escape from a horrible death was most marvellous, and in another instant he would have been beyond all human aid. Our efforts to sound the depths of this spring with a pole thirty-five feet in length were fruitless."

The report of the Geological Expedition describes these curious springs somewhat more minutely. The first that attracted Dr. Hayden's attention was the powerful steam-vent above mentioned, which he calls the Locomotive Jet. "The aperture is about six inches in diameter, a sort of raised chimney, and all around are numerous small continuous steam-vents, all of which are elegantly lined with the bright-yellow sulphur. The entire surface is covered with the white silicious crust, which gives forth a hollow sound beneath the tread; and we took pleasure in breaking it up in the vicinity of the vents, and exposing the wonderful beauty of the sulphur-coating on the inner sides. This crust is ever hot, and yet so firm that we could walk over it anywhere. On the south side of these hills, close to the foot, is a magnificent sulphur-spring. The deposits around it are silica; but some places are white, and enamelled like the finest porcelain. The thin edges of the nearly circular rim extend over the waters of the basin several feet, yet the open portion is fifteen feet in diameter. The water is in a constant state of agitation. The steam that issues from this spring is so strong and hot that it was only on the windward side that I could approach it and ascertain its temperature, 197°. The agitation seemed to affect the entire mass, carrying it up impulsively to the height of four feet. It may be compared to a huge caldron of perfectly clear water somewhat superheated. But it is the decorations about this spring that lent the charm, after our astonishment at the seething mass before us—the most beautiful scolloping around the rim, and the inner and outer surface covered with a sort of pearl-like bead work. The base is the pure white silica, while the sulphur gave every possible shade, from yellow to the most delicate cream. No kind of embroidering that human art can conceive or fashion could equal this specimen of the cunning skill of nature. On the northeast side of the hills, extending from their summits, are large numbers of the steam-vents, with the sulphur linings and deposits of the sulphur over the surface. These hills are entirely due to the old hot springs, and are from 50 to 150 feet in height. The rock is mostly compact silica, but there is almost every degree of purity, from a kind of basalt to the snow-white silica. Some of it is a real conglomerate, with a fine silicious cement inclosing pebbles of white silica, like those seen around the craters of some geysers. Although at the present time there are no true geysers in this group, the evidence is clear that these were, in former times, very powerful ones, that have built up mountains of silica by their overflow. The steam-vents on the side and at the foot of these hills represent the dying stages of this once most active group. Quite a dense growth of pines now covers these hills, which rise up in the midst of the plains, and from their peculiar white appearance are conspicuous for a great distance. At one point there is a steam-vent so hot that it is difficult to approach it, emitting a strong sulphurous smell, and within two feet of it there is a larger spring, boiling like a caldron. So far as I can determine, there is no underground connection of any of the springs with each other. Sometimes the rims of these craters, as well as the inner sides of their basins, have a beautiful papulose surface, the silica just covered with a thin veil of delicate creamy sulphur. At this locality are some very remarkable turbid and mud springs. One of them has a basin twenty feet in diameter, nearly circular in form, and the contents have almost the consistency of thick hasty-pudding. Indeed, there is no comparison that can bring before the mind a clearer picture of such a mud volcano than a huge caldron of thick mush. The surface is covered all over with puffs of mud, which, as they burst, give off a thudlike noise, and then fine mud-waves recede from the centre of the puffs in the most perfect rings to the side. Although there are hundreds of these mudpots, yet it is very rare that the mud is in just the condition to admit of these peculiar rings. The thud is, of course, produced by the escape of the sulphureted hydrogen gas through the mud. The mud is so fine as to have no visible or sensible grain, and is very strongly impregnated with alum. For three hundred yards in length and twenty-five yards in width, the valley of this little branch of Alum Creek is perforated with these mud-vents of all sizes, and the contents are of all degrees of consistency, from merely turbid water to a thick mortar. The entire surface is perfectly bare of vegetation, and hot, yielding in many places to a slight pressure. I attempted to walk about among these simmering vents, and broke through to my knees, covering myself with the hot mud, to my great pain and subsequent inconvenience. One of the largest of the turbid springs has a basin with a nearly circular rim twenty feet from the margin to the water, and forty feet in diameter. There are two or three centres of ebullition; temperature, 188°."

A couple of miles above these springs, near the banks of the Yellowstone, is a not less remarkable group of sulphur and mud springs. All the intermediate space abounds in the remains of similar springs, now quiescent or dead, yet giving evidence of former power and activity beyond that displayed by any now existing. "There were giants in those days!" Mr. Langford describes a group of these "unsightly caldrons," varying in size from two to ten feet in diameter; their surfaces from three to eight feet below the level of the plain: "The contents of the most of them were of the consistency of thick paint, which they greatly resembled, some being yellow, others pink, and others dark brown. This semi-fluid was boiling at a fearful rate, much after the fashion of a hasty-pudding in the last stages of completion. The bubbles, often two feet in height, would explode with a puff, emitting at each time a villainous smell of sulphuretted vapor. Springs six and eight feet in diameter, but four feet asunder, presented distinct phenomenal characteristics. There was no connection between them, above or below. The sediment varied in color, and not unfrequently there would be an inequality of five feet in their surfaces. Each, seemingly, was supplied with a separate force. They were embraced within a radius of 1,200 feet, which was covered with a strong incrustation, the various vents in which emitted streams of heated vapor. Our silver watches, and other metallic articles, assumed a dark leaden hue. The atmosphere was filled with sulphurous gases, and the river opposite our camp was impregnated with the mineral bases of adjacent springs. At the base of adjacent foot-hills we found three springs of boiling mud, the largest of which, forty feet in diameter, encircled by an elevated rim of solid tufa, resembles an immense caldron. The seething, bubbling contents, covered with steam, are five feet below the rim. The disgusting appearance of this spring is scarcely atoned for by the wonder with which it fills the beholder. The other two springs, much smaller, but presenting the same general features, are located near a large sulphur spring of milder temperature, but too hot for bathing. On the brow of an adjacent hillock, amid the green pines, heated vapor issues in scorching jets from several craters and fissures. Passing over the hill, we struck a small stream of perfectly transparent water flowing from a cavern, the roof of which tapers back to the water, which is boiling furiously, at a distance of twenty feet from the mouth, and is ejected through it in uniform jets of great force. The sides and entrance of the cavern are covered with soft, green sediment, which renders the rock on which it is deposited as soft and pliable as putty.

"About two hundred yards from this cave is a most singular phenomenon, which we called the Muddy Geyser. It presents a funnel-shaped orifice, in the midst of a basin one hundred and fifty feet in diameter, with sloping sides of clay and sand. The crater or orifice, at the surface, is thirty by fifty feet in diameter. It tapers quite uniformly to the depth of about thirty feet, where the water may be seen, when the geyser is in repose, presenting a surface of six or seven feet in breadth. The flow of this geyser is regular every six hours. The water rises gradually, commencing to boil when about half way to the surface, and occasionally breaking forth in great violence. When the crater is filled, it is expelled from it in a splashing, scattered mass, ten or fifteen feet in thickness, to the height of forty feet. The water is of a dark lead color, and deposits the substance it holds in solution in the form of miniature stalagmites upon the sides and top of the crater. As this was the first object which approached a geyser, we, naturally enough, regarded it with intense curiosity....

The Mud Volcano.

"While returning by a new route to our camp, dull, thundering sounds, which General Washburn likened to frequent discharges of a distant mortar, broke upon our ears. We followed their direction, and found them to proceed from a mud volcano, which occupied the slope of a small hill, embowered in a grove of pines. Dense volumes of steam shot into the air with each report, through a crater thirty feet in diameter. The reports, though irregular, occurred as often as every five seconds, and could be distinctly heard half a mile. Each alternate report shook the ground a distance of two hundred yards or more, and the massive jets of vapor which accompanied them burst forth like the smoke of burning gunpowder. It was impossible to stand on the edge of that side of the crater opposite the wind, and one of our party, Mr. Hedges, was rewarded for his temerity in venturing too near the rim, by being thrown by the force of the volume of steam violently down the outer side of the crater. From hasty views, afforded by occasional gusts of wind, we could see at a depth of sixty feet the regurgitating contents."

"This volcano, as is evident from the freshness of the vegetation and the particles of dried clay adhering to the topmost branches of the trees surrounding it, is of very recent formation. Probably it burst forth but a few months ago. Its first explosion must have been terrible. We saw limbs of trees 125 feet high encased in clay, and found its scattered contents two hundred feet from it."

On the east side of the Yellowstone, close to the margin of the river, are a few turbid springs, and mud-springs strongly impregnated with alum. The mud is yellow and contains much sulphur. These, the discoverers, Dr. Hayden and his company, called Mud-sulphur Springs. The main basin is 15 by 30 feet, and has three centres of ebullition, showing that deep in the earth are three independent orifices for the emission of heated waters. Dr. Hayden's description of the roaring spring issuing from a cavern, coincides with that given above. He called it the Grotto. Around all these springs he observed an abundance of grasses, rushes, mosses, and other plants growing with a surprising luxuriance. The recent mud-volcano described by Mr. Langford was considered by Dr. Hayden as the most remarkable mud-spring thus far discovered in the West.

"It does not boil with an impulse like most of the mud-springs," he says, "but with a constant roar which shakes the ground for a considerable distance, and may be heard for half a mile. A dense column of steam is ever rising, filling the crater, but now and then a passing breeze will remove it for a moment, revealing one of the most terrific sights one could well imagine. The contents are composed of thin mud in a continual state of the most

First flowing	9.20 A.M. to 9.35 A.M	Length	15 minutes
Second flowing	1.30 P.M. to 1.50 P.M.	Length	20 minutes
Third flowing	5 P.M. to 5.15 P.M.	Length	15 minutes
Fourth flowing	8.30 P.M. to 8.50 P.M.	Length	20 minutes
Fifth flowing	12.30 P.M. to 12.45 P.M.	Length	15 minutes
Sixth flowing	4 A.M. to 4.15 A.M.	Length	15 minutes
Seventh flowing	7.30 A.M. to 7.45 A.M.	Length	15 minutes
Eighth flowing	11 A.M. to 11.10 A.M.	Length	10 minutes
Total length of time			26 hours
Aggregate time of flowing			Three hours and 15 minutes
Average length of flowings			15 minutes and 37 and one half seconds

violent agitation, like an immense caldron of mush submitted to a constant, uniform, but most intense heat.... All the indications around this most remarkable caldron show that it has broken out at a recent period; that the caving in of the sides so choked up the orifice that it relieved itself, hurling the muddy contents over the living pines in the vicinity."

The steam rising from this spring—the Giant's Caldron—can be seen for many miles in every direction. The movements of Muddy Geyser were closely watched for twenty-four hours by Mr. Campbell Carrington, who was specially detailed for that duty by Dr. Hayden. His observations began about nine o'clock A.M., July 1st. Then the pool was calm. Shortly after, he heard the loud, hissing noise of escaping steam. Hurrying to the geyser, he saw a wave about three feet in height rise and die away to the left; three similar waves followed in quick succession. Their dense columns of steam burst up to the height of twenty feet, with a dull, heavy explosion, the action continuing for fifteen minutes, when the spring ceased flowing as suddenly as it had begun. The average height of the flowing was about fifteen feet, though some of the jets reached fully thirty feet. Five minutes after the eruption the pool measured twenty-five feet in circumference and three in depth, where before it was a hundred feet in circumference and eleven in depth. Ten minutes later the mud began to rise slowly in the pool. This continued for a little over three hours, when the spring began to boil near the centre. The ebullition gradually increased in violence for twenty minutes, then it suddenly stopped, and the eruption began as at first. This rising, falling, and overflowing took place eight times in twenty-four hours. The table on p.69 shows the time of the observed flowings and their length:

Chapter 10

YELLOWSTONE LAKE

"Such a vision," exclaims the sober-minded chief of the Geological Survey, "is worth a lifetime; and only *one* of such marvellous beauty will ever greet human eyes."

"Secluded amid the loftiest peaks of the Rocky Mountains," writes Mr. Langford, "possessing strange peculiarities of form and beauty, this watery solitude is one of the most attractive natural objects in the world. Its southern shore, indented with long narrow inlets, not unlike the frequent fiords of Iceland, bears testimony to the awful upheaval and tremendous force of the elements which resulted in its creation. The long pine-crowned promontories, stretching into it from the base of the hills, lend new and charming features to an aquatic scene full of novelty and splendor. Islands of emerald hue dot its surface, and a margin of sparkling sand forms its jewelled setting. The winds, compressed in their passage through the mountain gorges, lash it into a sea as terrible as the fretted ocean, covering it with foam. But now it lay before us calm and unruffled, save by the gentle wavelets which broke in murmurs along the shore. Water, one of the grandest elements of scenery, never seemed so beautiful before. It formed a fitting climax to all the wonders we had seen, and we gazed upon it for hours, entranced with its increasing attractions."

The beautiful sheet of water so enthusiastically yet fittingly described, is somewhat more than twenty miles long and fifteen broad, with an irregular outline, presenting some of the loveliest shore-lines that water ever assumed. Its form has been compared to that of an outspread hand, the northern portion representing the palm, the southwestern a swollen thumb, the first and second fingers aborted, the third and fourth disproportionately large. A glance at the map will show that a juster comparison would be to the head and shoulders of some grotesque animal with two slender ears and a pair of huge knobby horns—the head facing the north. The greatest stretch of water extends from the end of the heavy lower jaw (the outlet of the Yellowstone) to the top of the upper horn, where the Upper Yellowstone comes in; while the great body of the water lies between the forehead and the base of the shoulder. The superficial area of the lake is about three hundred square miles; its greatest depth 300 feet, and its elevation above the sea 7,427 feet. In the last respect it has but one rival, Lake Titicaca in South America.

Yellowstone Lake.

Lying upon the very crown of the continent, Yellowstone Lake receives no tributaries of any considerable size, its clear cold water coming solely from the snows that fall on the lofty mountain ranges that hem it in on every side. In the early part of the day, when the air is still and the bright sunshine falls on its unruffled surface, its bright green color, shading to a delicate ultramarine, commands the admiration of every beholder. Later in the day, when the mountain winds come down from their icy heights, it puts on an aspect more in accordance with the fierce wilderness around it. Its shores are paved with volcanic rocks, sometimes in masses, sometimes broken and worn into pebbles of trachyte, obsidian, chalcedony, cornelians, agates, and bits of agatized wood; and again, ground to obsidian-sand and sprinkled with crystals of California diamonds. Here and there hot-spring deposits show wave-worn bluffs of the purest white; and in sheltered bays, clay-concretions and casts from mud-puffs strew the beach with curious forms, that exploring trappers mistook for the drinking cups, stone war-clubs, and broken idols of some extinct race.

Vegetation is abundant in the lake as well as around it. Several species of plants grow far out into the deep waters, living thickly on rocks twenty feet below the surface. After a severe storm their uptorn stems strew the beach like kelp on the seashore, and the water is discolored with vegetable matter for several yards from the shore. The water swarms with trout, but there is no other kind of fish, no shells, no shell-fish,—nothing but trout. Of these, Mr. Carrington, the naturalist of the Geological Survey, reports the following interesting observations:

"Although I searched with diligence and care in the neighboring streams and waters around the Yellowstone Lake, I was unable to find any other species of fish except the salmon-trout; their numbers are almost inconceivable; average weight, one pound and a half; color, a light-grey above, passing into a light-yellow below; the fins, all except the dorsal and caudal, vary from a bright-yellow to a brilliant orange, they being a dark-grey and heavily spotted. A curious fact, and one well worthy of the closest attention of an aspiring icthyologist, is connected with these fish, namely, that among their intestines, and even interlaced in their solid flesh, are found intestinal worms, varying in size, length, and thickness, the largest measuring about six inches in length. On cutting one of these trout open, the first thing that attracts your attention is small oleaginous-looking spots clinging to the intestines, which, on being pressed between the fingers, break and change into one of these worms, small, it is true, but nevertheless perfect in its formation. From five or six up to forty or fifty will be found in a trout, varying, as I said before, in size, the larger ones being found in the solid flesh, through which they work their way, and which, in a very short while, becomes almost putrid. Their number can generally be estimated from the appearance of the fish itself; if many, the trout is extremely poor in flesh, the color changes from the healthy grey to a dull pale, it swims lazily near the top of the water, losing all its shyness and fear of man; it becomes almost savage in its appetite, biting voraciously at anything thrown in the water, and its flesh becomes soft and yielding. If, on the other hand, there are few or none, the flesh of the fish is plump and solid, and he is quick and sprightly in all his motions. I noticed that it was almost invariably the case when a trout had several scars on the outside of his body that it was free from these worms, and I therefore took it for granted that the worms finally worked their way through the body, and the flesh, on healing up, leaves the scars on the outside; the trout, in a short while, becomes plump and healthy again. The only way that I can account for the appearance of these worms is, that the fish swallows certain bugs or insects, and that the larvæ formed from them gradually develop into the full-grown intestinal worm. But even if this explanation of their appearance was received, does it not seem a little strange that while all the fish above the Upper Falls are more or less affected by them, that below and even between the Upper and Lower Falls such a thing as wormy trout is never heard of? Being unable, with my limited knowledge of icthyology, to arrive at any definite conclusion in regard to their appearance, I submit the above facts to those who are more learned than myself in this most interesting branch of natural history."

Waterfowl make up in number and variety for the lack of life within the lake. The surface fairly swarms with them. Lieutenant Doane enumerates swans, pelicans, gulls, geese, brants, and many varieties of ducks and dippers; also herons and sand-hill cranes. The pelicans are very plentiful, immense fleets of them sailing in company with the majestic swan, and at nightfall the low, flat islands in the lake are white with them. The gulls are of the same variety as those of San Francisco Harbor. Eagles, hawks, ravens, ospreys, prairie chickens, grouse, mocking-birds and woodpeckers are common in and around the lake basin. Mention is also made of a guide-bird, whose habits correspond with its name. It resembles the blackbird, but is larger. Lieutenant Doane says:

"I saw but one of these—the day I went to the bottom of the Grand Cañon; it hopped and flew along from rock to rock ahead of us during the whole trip down, waited perched upon a rock while we were resting, and led us clear to the summit again in the same manner, making innumerable sounds and gestures constantly to attract attention. Others of the party remarked birds of the same kind and acting in the same manner."

Herds of deer, elk, and mountain sheep, throng the forests and mountain meadows about the lake. Buffalo signs, grizzly bears and California lions are far from uncommon, while the smaller lakes and creek-valleys of the basin are fairly alive with otter, beaver, mink, and muskrats. Lieutenant Doane observed several unnamed and undescribed species of squirrels and weasels, and doubtless there are many other new varieties of animal life peculiar to this little-known region. One department of natural history, however, is happily unrepresented in the basin. There are no snakes, though rattlesnakes are plentiful down the Yellowstone.

There are but two considerable islands in the lake—Stephenson's and Frank's—each about a mile long, narrow and covered with a thick growth of pines. Dot Island, near Frank's, a small lozenge-shaped mud-bank, not over a third of a mile long, and half a dozen of smaller size, usually near the shore, complete the list.

The first explorers constructed a rude raft for the purpose of visiting these islands and exploring the shore-line of the lake, but it was speedily wrecked by the choppy waves beat up by the sudden gusts from the mountains. The Geological Expedition took the precaution to carry from Fort Ellis the framework of a little craft, twelve feet long, three and a half feet wide, and twenty-two inches deep, which, covered with well-tarred canvas, made a very serviceable boat for fair-weather navigation. "Our little bark, whose keel was the first to plow the waters of the most beautiful lake on the continent," says Dr. Hayden, "was named by Mr. Stephenson in compliment to Miss Anna L. Dawes, the amiable daughter of Hon. H. L. Dawes. My whole party," he adds, "were glad to manifest, by this slight tribute, their gratitude to the distinguished statesman, whose generous sympathy and aid had contributed so much toward securing the appropriation which enabled them to explore this marvellous region."

The First Boat on Yellowstone Lake.

The little craft rode the waves well and performed excellent service. Its first voyage was to Stephenson's Island, named after the first assistant of the expedition.

action at some period not very remote, and that the steam-vents, which are very numerous, represent only the dying stages. These vents or chimneys are richly adorned with brilliant yellow sulphur, sometimes as a hard amorphous coating, and sometimes in delicate crystals that vanish like frost-work at the touch. It seems that it is only during the last stages of these springs that they adorn themselves with these brilliant and vivid colors.

Hot springs are scattered along the valley of the creek for several miles, some of them of considerable size and beauty. The average width of the valley is about two miles; the heat from the springs and the extremely fertile soil combining to fill the valley with abundant vegetation. At the northeastern corner of the lake, five or six miles from the outlet, is a long, low spit of land built out into the lake by ancient geyser action. A few roaring steam-vents, giving name to the point, are all that remain of the violent action that once characterized the place. The hot spring area is four or five miles long by two wide; the ground in many places being perforated like a cullender with simmering vents. A mile or so from the lake is a large pond where there is another extensive group of springs, depositing sulphur, alum, common salt, and staining the ground with oxide of iron.

South of Steam Point is a small bay bounded by a deposit of yellow clay, full of the remarkable concretions already referred to. Further up the eastern shore are pebbly beaches strewn with agates cornelians, and chips of chalcedony. Beyond, the narrow lake-shore is quite impassable. The adjacent lowlands, and the higher levels and hill-slopes further back, are almost as difficult of penetration, owing to the dense growth of lofty pines and the interminable fire-slashes that cover large areas. These fire-slashes are due to autumnal fires which sweep through the forests, burning the vegetable mould, so that the trees are left without support, and the first wind lays them down in the wildest confusion. Through these networks of fallen timber it is with the utmost difficulty that a passage can be forced. All the explorers speak of the exasperating nature of their tribulations in these wildernesses.

Mr. Langford treats it with characteristic good humor.

"Ascending the plateau from the beach," he says, "we became at once involved in all the intricacies of a primeval wilderness of pines. Difficulties

Chapter 11

AROUND YELLOWSTONE LAKE

The Yellowstone leaves the Lake with an easy flow in a channel a quarter of a mile wide, and deep enough to swim a horse. A mile to the eastward of the outlet is the mouth of Pelican Creek, whose swampy valley is the resort of myriads of waterfowl. On the northern side, three or four miles from the lake, Sulphur Hills stand as monuments of a once magnificent system of boiling springs.

The deposit covers the side of the mountain to an elevation of 600 feet above the lake shore. The huge white mass of silica, covering an area half a mile square, can be seen from any position on the lake shore, whence it appears like an immense bank of snow. In the valley near Pelican Creek, a few springs issue from beneath the crust, distributing their waters over the bottom and depositing oxide of iron, sulphur, and silica in the most beautiful blending of gay colors. Although the waters of the springs are 160° in temperature, the channels are lined with a thick growth of mosses and other plants, and in the water is an abundance of vividly green vegetation. The mass of hot-spring material built up here cannot be less than 400 feet in thickness. A large portion of it is pudding-stone or conglomerate. Some of the masses inclosed in the fine white silicious cement are themselves globes of pure white silica, eight inches in diameter. It is plain, from the evidence still remaining, that this old ruin has been the theatre of tremendous geyser increased with our progress through it, severely trying the amiability of every member of the company. Our pack-horses would frequently get wedged between the trees or caught in the traps of a network of fallen trunks, from which labor, patience, and ingenuity were severely taxed to extricate them. The ludicrous sometimes came to our relief, proving that there was nothing so effectual in allaying excitement as hearty laughter. We had a remarkable pony in our pack-train, which, from the moment we entered the forest, by his numerous acrobatic performances and mishaps furnished amusement for the company. One part of the process of travel through this forest could only be accomplished by leaping over the fallen trunks, an exploit which, with all the spirit needful for the purpose, our little broncho lacked the power always to perform. As a consequence, he was frequently found with the feat half accomplished, resting upon the midriff, his fore and hind feet suspended over the opposite sides of some huge log. His ambition to excel was only equalled by the patience he exhibited in difficulty. On one occasion, while clambering a steep rocky ascent, his head overtopping his haunches, he literally performed three of the most wonderful backward headsprings ever recorded in equine history. A continued experience of this kind, after three weeks' toilsome travel, found him as sound as on the day of its commencement, and we dubbed him the 'Little Invulnerable.'"

In another place Mr. Langford writes:

"Our journey of five miles, the next day, was accomplished with great difficulty and annoyance. Almost the entire distance was through a forest piled full of fallen trunks. Travelling was but another name for scrambling; and as man is at times the least amiable of animals, our tempers frequently displayed alarming activity, not only towards the patient creatures laden with our stores, but towards each other. Once, while involved in the reticulated meshes of a vast net of branches and tree-tops, each man, with varied expletive emphasis, clamorously insisting upon a particular mode of extrication, a member of the party, who was always jolly, restored us to instant good-humor by repeating, in theatrical tone and manner, those beautiful lines from Childe Harold:—

> "There is a pleasure in the pathless woods,

There is a rapture on the lonely shore."

Our 'Little Invulnerable,' too, was the unconscious cause of many bursts of laughter, which, like the plaudits of an appreciative audience, came in at the right time."

The eastern rim of the Yellowstone Basin is formed by one of the grandest volcanic ranges in the world, the general level of their summits being about 10,000 feet above the sea, while numerous peaks thrust their rugged crests a thousand feet higher into the sky. Mr. Langford and Lieutenant Doane were the first to penetrate this range, climbing with great labor one of the highest of the groups of lofty peaks near the southeast corner of the Lake.

"The grandeur and vast extent of the view from this elevation," writes Mr. Langford, "beggar description. The lake and valley surrounding it lay seemingly at our feet within jumping distance. Beyond them we saw with great distinctness the jets of the mud volcano and geyser. But beyond all these, stretching away into a horizon of cloud-defined mountains, was the entire Wind River range, revealing in the sunlight the dark recesses, gloomy cañons, stupendous precipices, and glancing pinnacles, which everywhere dotted its jagged slopes. Lofty peaks shot up in gigantic spires from the main body of the range, glittering in the sunbeams like solid crystal. The mountain on which we stood was the most westerly peak of a range which, in long-extended volume, swept to the southeastern horizon, exhibiting a continuous elevation more than thirty miles in width; its central line broken into countless points, knobs, glens, and defiles, all on the most colossal scale of grandeur and magnificence. Outside of these, on either border, along the entire range, lofty peaks rose at intervals, seemingly vying with each other in the varied splendors they presented to the beholder. The scene was full of majesty. The valley at the base of this range was dotted with small lakes and cloven centrally by the river, which, in the far distance, we could see emerging from a cañon of immense dimensions, within the shade of which two enormous jets of steam shot to an incredible height into the atmosphere."

Between the lake and this group of mountains—the three highest of which bear the names of Langford, Doane, and Stephenson—is Brimstone

Basin. For several miles the ground is impregnated with sulphur, and the air is tainted with sulphurous exhalations. Streams of warm sulphur-water course the hillsides and unite to form a considerable rivulet called Alum Creek, whose channel is coated with a creamy-white mixture of silica and sulphur. Old pine logs, once lofty trees, lie prostrate in every direction over the basin, which covers an area some three miles in extent. From all appearances this basin must have been the scene of thermal activity within a comparatively recent period; but now not a spring can be found with a temperature above that of ordinary spring-water. Similar brimstone basins are numerous around the lake, on the lower slopes of the mountains, at the foot of bluffs, or more frequently in level districts. The latter are always wet, and generally impassable, the thin crust covering an abundance of scalding mud, especially dangerous to horses.

The Upper Yellowstone rises in the high volcanic range which shuts off the Yellowstone Basin from the Wind River drainage, forming what is known as the great water-shed of the continent.

This range of mountains has a marvellous history. As it is the loftiest, so it is the most remarkable lateral ridge of the Rocky Range. The Indians regard it as the "crest of the world," and among the Blackfeet there is a fable that he who attains its summit catches a view of the land of souls, and beholds the happy hunting-grounds spread out below him, brightening with the abodes of free and generous spirits.

In the expedition sent across the continent by Mr. Astor, in 1811, under command of Captain Wilson P. Hunt, that gentleman met with the first serious obstacle to his progress at the base of this range. After numerous efforts to scale it, he turned away and followed the valley of the Snake, encountering the most discouraging disasters until he arrived at Astoria.

Later, in 1833, the indomitable Captain Bonneville was lost in this mountain labyrinth, and, after devising various modes of escape, finally determined to ascend the range, which tremendous task he succeeded in accomplishing, in company with one of his men. It was this same line of snow-clad, craggy peaks that turned back Captain Raynolds in 1859.

Breaking Through.

Near its mouth the Upper Yellowstone is about half the size of the main stream as it leaves the lake. Its valley is about three miles wide and very marshy; all the little streams flowing down from the wooded hill-slopes being obstructed by beaver-dams, so as to form continuous chains of ponds. The sides of the valley are dark, sombre walls of volcanic rock, which weathers into curious and imposing forms. Looking up the valley from some high point, one almost imagines himself in the presence of the ruins of some gigantic city, so much like ancient castles and cathedrals do these rocks appear—a deception that is not a little heightened by the singular vertical furrows cut deep into the cliffs. At the base of the walls immense masses of breccia have fallen from the mountain tops, in many instances cutting long swaths through the pine forests. In the upper part of the valley, which in midsummer is lush with vegetation, five streams flow down from the mountains to swell the waters of the Yellowstone. These streams Colonel Barlow calls, in honor of his commander's greatest victory, the Five Forks.

Here the valley terminates abruptly, the mountains rising like walls and shutting off the country beyond. Just at the head of the valley is a little lake, a hundred yards or so in width; the large lake which has been placed on maps as Bridger's Lake having no existence. Dr. Hayden with two assistants ascended the mountains to the west of the head of the valley to survey the district bordering on the great divide. From this point as far as the eye can reach on every side are bare, bald peaks, domes and ridges in great numbers. At least one hundred peaks worthy of a name can be located within the radius of vision. The rocks everywhere, though massive, black, and deeply furrowed vertically, have the appearance of horizontal stratification. In some instances the furrows are so regular that the breccia has a columnar appearance. The summits of the mountains are composed entirely of breccia, containing angular masses of trachyte, from 10 to 30 feet in diameter, though most of the fragments are small. Dr. Hayden's party camped at night near a small lake, by the side of a bank of snow, 10,000 feet above the sea, with short spring grass and flowers all around them. There are but two seasons on these mountain summits, spring and winter; as late as August fresh new grass may be seen springing up where a huge bank of snow has just disappeared. Little spring-flowers, seldom more than two or three inches high, cover the ground—*Clatonia*, *Viola*, *Ranunculus*, and many others. The following morning they travelled for several miles along a ridge not more than two hundred yards wide, from one side of which the waters flow into the Pacific, and on the other, into the Atlantic. To the westward the outlines of the Teton Range, with its shark-teeth summits, are most clearly visible, covered with snow. From whatever point of view, the sharp-pointed peaks of this range have the form of huge sharks' teeth. To the southward, for fifty miles at least, nothing but igneous rocks can be seen. Toward the Tetons there is a series of high ridges, passing off from the main Teton Range toward the northeast, and varying in height from 9,500 to 10,500 feet above the sea, and from 1,000 to 1,800 feet above the valleys at their base.

The explorers ascended one of the high ridges, (not the highest, however,) and found it to be 1,650 feet above the valley at its foot. The northeast side is steep like a roof, the southwest breaking off abruptly. From the summit of this ridge, the view is grand in the extreme. To the westward the entire country, for the distance of fifty miles, seems to have been thrown up into high, sharp ridges, with gorges 1,000 to 1,500 feet in depth. Beautiful lakes, grassy meadows also, come within the field of vision. "I can conceive," says Dr. Hayden, "of no more wonderful and attractive region for the explorer. It would not be difficult for the traveller to make his way among these grand gorges, penetrating every valley, and ascending every mountain and ridge. The best of grass, wood, water and game are abundant to supply the wants of himself and animals.

"I think," he continues, "that numerous passes could be found from the valley of Snake River to the basin of the Yellowstone. It seems to me there are many points on the south rim of the basin where a road could be made with ease into the valley of Snake River. From this ridge there appears to be but little difference in the altitude of Yellowstone Lake and Heart Lake, and they cannot be more than eight or ten miles apart, and yet the latter is one of the sources of Snake River. The little branches of Snake River nearly interlace with some streams that flow into the lake, and the gullies come up within two miles of the shore-line. There is a very narrow dividing ridge in one place, between the drainage, which may be within one mile of the lake."

Heart Lake was visited by Colonel Barlow, who found it a pretty, pear-shaped sheet of water, four miles long and two wide in its broadest part. From the north it receives a warm creek fed by a considerable group of hot springs. Its outlet at the southern end joins the terminal creek of Snake River, a few miles from its source among the Yellowstone Mountains.

Ten miles northwest of Heart Lake is Madison Lake, the source of Madison River, the country between being a somewhat rugged range of mountains, of which Red Mountain is the most conspicuous. To the eastward from Heart Lake is Mount Sheridan, from the summit of which a magnificent view of the Yellowstone Basin can be obtained. Nearer the great lake is Flat Mountain, whose altitude falls a little short of 10,000 feet. Between Flat Mountain and the Yellowstone Range the divide is very low, some of the branches of Snake River extending up to within two miles of the lake, where the elevation is not more than 400 feet above the lake level. It is doubtless this singular interlacing of the headwaters of the Yellowstone and Snake River that gave rise to Bridger's story of the "Two Ocean River."

At sunrise on the morning of August 10th, at the west base of Flat Mountain, the thermometer stood at 15½° Fah., and water froze in Dr. Hayden's tent that night a quarter of an inch thick. It was in this neighborhood that Mr. Everts was lost from the first expedition.

The country between Flat Mountain and the hot springs at the southwestern extremity of the lake is a level plateau with alternating spaces of grassy glade and dense thickets of pine around and between a perfect network of small, lily-covered lakes. The hot springs on the lake shore are numerous and of great variety and interest. There are no true geysers, however, though some of the springs are pulsating springs, the water rising and falling in their orifices with great regularity. Higher up the bank are a large number of mud-springs, two or three hundred in all, of variable temperatures, the most of them not differing materially from those already described. Some, however, have a character strikingly unique. The area covered by the springs is about three miles long and half a mile wide, a portion of it reaching out into the lake. Some of the submerged springs have built up funnel-shaped craters of silicious deposit, from five to twenty feet in height, rising from the bottom to the surface of the water. Extending a pole over the deep water, members of Dr. Hayden's party caught trout and cooked them in these boiling springs out in the lake without removing them from the hook.

Four hundred yards from the lake shore is a large boiling basin of pink-colored mud, seventy feet in diameter, with a rim of conical mud craters, which project the hot mud in every direction. The deposit speedily hardens into a firm, laminated stone, of beautiful texture, though the brilliant pink color fades to a chalky white. Near and around this basin are a dozen springs, from six to twenty-five feet across, boiling muddy water of a paint-like consistency, varying in color from pure white to dark yellow. Close by are several flowing springs of clear hot water, from ten to fifty feet in diameter, their basins and channels lined with deposits of red, green, yellow, and black, giving them an appearance of gorgeous splendor. The bright colors are on the surface of the rock only, which is too friable to be preserved. Below these springs are several large craters of bluish water, boiling to the height of two feet in the centre, and discharging large streams of water; their rims are raised a few inches in a delicate rock-margin of a fringe-like appearance, deposited from the water. Beyond these are two lakes of purple water, hot, but not boiling, and giving deposits of great beauty. Nearby are two more blue springs, one thirty by forty feet, and 173° in temperature. This spring discharges a considerable stream into the other, which is seventy feet distant, and six feet lower. The latter is forty feet by seventy-five, 183° in temperature, and discharges a stream of one hundred inches. The craters of these springs are lined with a silvery-white deposit of silica, which reflect the light so as to illuminate the water to an immense depth. Both craters have perpendicular but irregular walls, and the distance to which objects are visible down in their deep abysses is truly wonderful.

West of these is another group of clear watered hot springs, which surpass all the rest in singularity if not in beauty. These have basins of different sizes and immeasurable depth, in which float what appear like raw bullock hides as they look in a tanner's vat, waving sluggishly with every undulation of the water. On examination, this leathery substance proves to be of fragile texture, like the vegetable scum of stagnant pools, and brilliantly colored red, yellow, green, etc., black on the underside. This singular substance is about two inches in thickness, jelly-like to the touch, and is composed largely of vegetable matter, which Dr. Hayden thinks to be diatoms.

Of the beautiful transparency of the springs above described, Dr. Hayden says: "So clear was the water that the smallest object could be seen on the sides of the basin; and as the breeze swept across the surface, the ultramarine hue of the transparent depth in the bright sunlight was the most dazzlingly beautiful sight I ever beheld. There were a number of these large clear springs, but not more than two or three that exhibited all those brilliant shades, from deep sea green to ultramarine."

Occasionally, says Lieutenant Doane, this anomaly is seen, namely: "two springs, at different levels, both boiling violently; one pours a large and constant stream into the other, yet the former does not diminish, nor does the latter fill up and overflow."

Most of the springs, however, seem to be independent of each other, since they have different levels at the surface, different temperatures and pulsations, and rarely are the waters and deposits of any two exactly alike.

Passing northward through dense woods and almost impenetrable fire-slashes, the next noteworthy region arrived at is the valley of Bridge Creek, the creek receiving its name from a natural bridge of trachyte thrown across the stream. The bridge is narrow, affording scanty room for the well-worn elk-trail two feet wide, while the descent on either hand is so great that a fall from the bridge would be fatal to man or beast. Numerous herds of elk make daily use of this convenient passway.

Dead and dying springs are abundant all along the valley of this creek, the most of them being reduced to mere steam-vents. In one place the spring deposits cover several acres and present a most attractive picture. The ground is thickly covered with conical mounds, from a few inches in diameter to a hundred feet, full of steaming orifices lined with brillant sulphur-crystals. The underside of the heated crust is everywhere adorned in the same manner. The basis of the deposit is snow-white silica, but it is variegated with every shade of yellow from sulphur, and with scarlet from oxide of iron. From a distance the whole region has the appearance of a vast lime-kiln in full operation. Most of the country has been eroded into rounded hills from fifty to two hundred feet high, composed of the whitish-yellow and pinkish clays and sands of the modern lake deposit, which seems to prevail more or less all-round the rim of the basin, reaching several hundred feet above the present level of the lake.

Between Bridge Creek and the outlet of the lake, completing the circuit of the basin, is the Elephant's Back, a long, low mountain, noticeable only for its rounded summit and precipitous sides.

Chapter 12

Upper Geyser Basin of Firehole River

Just over the western margin of the Yellowstone Basin, yet within the limits of our great National Park, is the grand geyser region of Firehole River. Here, in a valley a dozen miles long and two or three wide, is an exhibition of boiling and spouting springs on a scale so stupendous that if all the corresponding phenomena of all the rest of the world could be brought into an equal area the display would seem as nothing in comparison.

Firehole River, the main fork of the Madison, has its source in Madison Lake, a beautiful sheet of water set like a gem among the mountains, dense forests of pines coming down to the very shores. A pointed ridge extends into the lake on the west side about half a mile, giving it the form of a heart. Its area is about three miles from north to south, and two from east to west. Its shores are paved with masses of trachyte and obsidian. The high mountains about the lake and along the river are gashed with deep gorges, with steep and jagged sides. Pines grow upon the mountain-sides where the declivity is so great that they cannot be scaled. In the obstructed gorges and on the mountain-tops, from 9,000 to 10,000 feet above the sea, little lakes occur every mile or so, nestled among the pines. Clear-watered mountain-torrents tumble down the almost vertical ridges to swell the Firehole, making cascades that in any other region would enjoy world-wide fame. Just before reaching the geyser-basin, some ten miles below the lake, the river roars through a deep gorge in the trachyte rock, and as it emerges, dashes over two cliffs, one twenty, the other fifty feet in height. "These pretty falls," writes Lieutenant Doane, "if located on an Eastern stream, would be celebrated in history and song; here, amid objects so grand as to strain conception and stagger belief, they were passed without a halt."

Shortly after, the cañon widens and the dominion of the Fire King begins. Scattered along both banks of the river are boiling springs from two to twelve feet across, all in active eruption. The craters of these springs are from three to forty feet high. Like the springs on Gardiner's River, these gradually seal themselves up by depositing mineral matter around and over their orifices. Numbers of such self-extinguished craters, now cones of solid rock, are scattered along the river-side. Two miles further down the stream is the upper geyser-basin, an open, rolling valley, two miles wide and three long, the mountains on either side rising 1,500 feet above the valley, with steep, heavily-timbered ledges of dark rock.

Hurrying down the Firehole, thinking the wonders of the Yellowstone country had been left behind, and anxious only to reach the settlements of the Madison Valley, the expedition of 1871 was startled and astonished to see at no great distance an immense volume of clear, sparkling water projected into the air to the height of one hundred and twenty-five feet. "Geysers! geysers!" exclaimed one of the company, and, spurring their jaded horses, they were soon gathered around an unexpected phenomenon—a perfect geyser. The aperture through which the column of water was projected was an irregular oval, three feet by seven in diameter. The margin of sinter was curiously piled up, the exterior crust filled with little hollows full of water, in which were globules of sediment, gathered around bits of wood and other nuclei. This geyser stands on a mound, thirty feet above the level of the surrounding plain, its crater rising five or six feet higher. It spouted at regular intervals nine times during the explorers' stay, the columns of boiling water being thrown from ninety to one hundred and twenty-five feet at each discharge, which lasted from fifteen to twenty minutes. They gave it the name of "Old Faithful."

"Near the crater, and as far as the irruptive waters reach," writes Lieutenant Doane, "the character of the deposit is very peculiar. Close around the opening are built up walls, eight feet in height, of spherical nodules, from six inches to three feet in diameter. These stony spheres, in turn, are covered with minute globules of stalagmite, incrusted with a thin glazing of silica. The rock, at a distance, appears the color of ashes of roses, but near at hand shows a metallic gray, with pink and yellow margins of the utmost delicacy. Being constantly wet, the colors are brilliant beyond description. Sloping gently from this rim of the crater in every direction the rocks are full of cavities in successive terraces, forming little pools, with margins of silica the color of silver, the cavities being of irregular shape, constantly full of hot water, and precipitating delicate, coral-like beads of a bright saffron. These cavities are also fringed with rock around the edges, in meshes as delicate as the finest lace. Diminutive yellow columns rise from their depths, capped with small tablets of rock, and resembling flowers growing in the water. Some of them are filled with oval pebbles of a brilliant white color, and others with a yellow frost-work which builds up gradually in solid stalagmites. Receding still farther from the crater, the cavities become gradually larger, and the water cooler, causing changes in the brilliant colorings, and also in the formations of the deposits. These become calcareous spar, of a white or slate color, and occasionally variegated. The water of the geyser is colorless, tasteless, and without odor. The deposits are apparently as delicate as the down on the butterfly's wing, both in texture and coloring, yet are firm and solid beneath the tread. Those who have seen the stage representations of "Aladdin's Cave," and the "Home of the Dragon Fly," as produced in a first-class theatre, can form an idea of the wonderful coloring, but not of the intricate frost-work, of this fairy-like, yet solid mound of rock, growing up amid clouds of steam and showers of boiling water. One instinctively touches the hot ledges with his hands, and sounds with a stick the depths of the cavities in the slope, in utter doubt in the evidence of his own eyes. The beauty of the scene takes away one's breath. It is overpowering, transcending the visions of the Moslem's Paradise."

As the next party of explorers were leaving the basin, ascending the river, this grand old geyser, which stands sentinel at the head of the valley, gave them a magnificent parting display. "With little or no preliminary warning," writes Dr. Hayden, "it shot up a column of water about six feet in diameter to the height of 100 to 150 feet, and by a succession of impulses seemed to hold it up steadily for the space of fifteen minutes, the great mass of water falling directly back into the basin, and flowing over the edges and down the sides in large streams. When the action ceases, the water recedes beyond sight, and nothing is heard but the occasional escape of steam until another exhibition occurs. This is one of the most accommodating geysers in the basin, and during our stay played once an hour quite regularly."

Old Faithful stands alone, though surrounded by a number of old geyser hills, whether built up in past ages by one spring shifting its position from time to time, or by a group of springs, now almost exhausted, it is impossible to tell.

Just across the river, and close to the margin, stands a silicious cone, very symmetrical, slightly corrugated on its exterior surface, three feet in height and five in diameter at its base. Its orifice is oval, with scalloped edges, and two feet by three in diameter. Of this unpretending cone Mr. Langford writes:

"Not one of our company supposed that it was a geyser; and among so many wonders it had almost escaped notice. While we were at breakfast upon the morning of our departure a column of water, entirely filling the crater, shot from it, which, by accurate triangular measurement, we found to be 219 feet in height. The stream did not deflect more than four or five degrees from a vertical line, and the eruption lasted eighteen minutes. We named it 'The Bee-hive.'"

A hundred yards further from the river, near the centre of the large group of spouting and boiling geysers, of which the Bee-hive is one, is a large oval aperture with scalloped edges, the diameters of which were eighteen and twenty-five feet, the sides corrugated and covered with a greyish-white silicious deposit, which was distinctly visible at the depth of one hundred feet below the surface.

"No water could be discovered," writes Mr. Langford, on his first approach to the spring, "but we could distinctly hear it gurgling and boiling at a great distance below. Suddenly it began to rise, boiling and spluttering, and sending out huge masses of steam, causing a general stampede of our company, driving us some distance from our point of observation. When within about forty feet of the surface it became stationary, and we returned to look down upon it. It was foaming and surging at a terrible rate, occasionally emitting small jets of hot water nearly to the mouth of the orifice. All at once it seemed seized with a fearful spasm, and rose with incredible rapidity, hardly affording us time to flee to a safe distance, when it burst from the orifice with terrific momentum, rising in a column the full size of this immense aperture to the height of sixty feet; and through and out of the apex of this vast aqueous mass, five or six lesser jets or round columns of water, varying in size from six to fifteen inches in diameter, were projected to the marvellous height of two hundred and fifty feet. These lesser jets, so much higher than the main column, and shooting through it, doubtless proceed from auxiliary pipes leading into the principal orifice near the bottom, where the explosive force is greater.... This grand eruption continued for twenty minutes, and was the most magnificent sight we ever witnessed. We were standing on the side of the geyser nearest the sun, the gleams of which filled the sparkling column of water and spray with myriads of rainbows, whose arches were constantly changing,—dipping and fluttering hither and thither and disappearing only to be succeeded by others, again and again, amid the aqueous column, while the minute globules into which the spent jets were diffused when falling sparkled like a shower of diamonds, and around every shadow which the denser clouds of vapor, interrupting the sun's rays, cast upon the column, could be seen a luminous circle radiant with all the colors of the prism, and resembling the halo of glory represented in paintings as encircling the head of Divinity. All that we had previously witnessed seemed tame in comparison with the perfect grandeur and beauty of this display. Two of these wonderful eruptions occurred during the twenty-two hours we remained in the valley. This geyser we named 'The Giantess.'"

The central spring of this group stands on the summit of a great mound built up in thin layers, by the continual but moderate overflow of the spring. The crater is twenty feet in diameter, slightly bubbling or boiling near the centre, and with a thin, elegant ring projecting a few inches over the water. Looking down into the clear water of this spring one seems to be gazing into fathomless depths, while the bright blue of the water is unequalled even by the sea. There are many such central springs, usually crowning the summits of mounds, and with projecting rims carved with an intricate delicacy truly marvellous, and adorned with colors that defy description. "The great beauty of the prismatic colors," writes Dr. Hayden, "depends much on the sunlight; about the middle of the day, when the bright rays descend nearly vertically, and a slight breeze just makes a ripple on the surface, the colors exceed comparison; when the surface is calm there is one vast chaos of colors, dancing, as it were, like the colors of a kaleidoscope. As seen through this marvellous play of colors, the decorations on the sides of the basin are lighted up with a wild, weird beauty, which wafts one at once into the land of enchantment; all the brilliant feats of fairies and genii in the Arabian Nights' Entertainments are forgotten in the actual presence of such marvellous beauty; life becomes a privilege and a blessing after one has seen and thoroughly felt its cunning skill."

Across the river, and a short distance below this group, is the largest and most imposing formation in the valley—the crater of Castle Geyser. This geyser receives its name from its resemblance to the ruins of an old castle as one enters the valley from the east. The deposited silica has crystallized in immense globular masses, like cauliflowers or spongiform corals, apparently formed about a nucleus at right angles to the centre. The entire mound is about forty feet high, and the chimney twenty feet. The lower portion rises in steps formed of thin laminæ of silica, mostly very thin, but sometimes compact, an inch or two thick. On the southeast side, where the water is thrown out continually, these steps are ornamented with the usual bead and shell work, with the large cauliflower-like masses: but the other portions are fast going to decay, and the débris are abundant. This has undoubtedly been one of the most active and powerful geysers in the basin; it still keeps up a great roaring inside, and every few moments, as observed by Dr. Hayden, it throws out a column of water to the height of ten or fifteen feet. Occasionally it seems to have more imposing eruptions, since on one occasion Lieutenant Doane saw it throw a column of water to the height of sixty feet, with the escape of heavy volumes of steam. The next year Colonel Barlow saw a similar display. According to the latter observer, the base of the crater is three hundred and twenty-five feet in circumference, and the turret one hundred and twenty-five. At the base of the turret lies a large petrified pine log, covered with a brilliant incrustation several inches thick.

Across the river, and a little below the Castle, are some fifty springs and geysers, the chief of which has been called Grand Geyser, its power seeming greater than that of any other in the valley. Lieutenant Doane describes this magnificent geyser as follows:

"Opposite camp, on the other side of the river, is a high ledge of stalagmite, sloping from the base of the mountain down to the river. Numerous small knolls are scattered over its surface; the craters of boiling springs from 15 to 25 feet in diameter; some of these throw water to the height of three and four feet. On the summit of this bank of rock is the grand geyser of the world, a well in the strata, 20 by 25 feet in diametric measurements, (the perceptible elevation of the rim being but a few inches,) and when quiet having a visible depth of 100 feet. The edge of the basin is bounded by a heavy fringe of rock, and stalagmite in solid layers is deposited by the overflowing waters. When an eruption is about to occur the basin gradually fills with boiling water to within a few feet of the surface, then suddenly, with heavy concussions, immense clouds of steam rise to the height of 500 feet, and the whole great body of water, 20 by 25 feet, ascends in one gigantic column to the height of 90 feet; from the apex of this column five great jets shoot up, radiating slightly from each other, to the unparalleled altitude of 250 feet from the ground. The earth trembles under the descending deluge from this vast fountain; a thousand hissing sounds are heard in the air; rainbows encircle the summits of the jets with a halo of celestial glory. The falling water plows up and bears away the shelly strata, and a seething flood pours down the slope and into the river.

It is the grandest, the most majestic, and most terrible fountain in the world. After playing thus for twenty minutes it gradually subsides, the water lowers into the crater out of sight, the steam ceases to escape, and all is quiet. This grand geyser played three times in the afternoon, but appears to be irregular in its periods, as we did not see it in eruption again while in the

The Grand Geyser of the Firehole Basin.

valley. Its waters are of a deep ultramarine color, clear and beautiful. The waving to and fro of the gigantic fountain, in a bright sunlight, when its jets are at their highest, affords a spectacle of wonder of which any description can give but a feeble idea. Our whole party were wild with enthusiasm; many declared it was 300 feet in height; but I have kept, in the figures as set down above, within the limits of absolute certainty." Dr. Hayden describes it with equal enthusiasm.

"We camped the evening of August 5th, in the middle of the Upper Geyser Basin, in the midst of some of the grandest geysers in the world. Colonel Barlow and Captain Heap, of the United States Engineers, were camped on the opposite side of the Firehole. Soon after reaching camp a tremendous rumbling was heard, shaking the ground in every direction, and soon a column of steam burst forth from a crater near the edge of the east side of the river. Following the steam by a succession of impulses, a column of water, apparently six feet in diameter, rose to the height of 200 feet, while the steam ascended a thousand feet or more. So steady and uniform did the force act that the column of water appeared to be held there for some minutes, returning into the basin in millions of prismatic drops. This was continued for about fifteen minutes, and the rumbling and confusion attending it could be compared only to that of a charge in battle. It would be difficult to describe the intense excitement which attended such a display. It is probable that if we could have remained in the valley several days, and become accustomed to all the preliminary warnings, the excitement would have ceased, and we could have admired calmly the marvellous ease and beauty with which this column of hot water was held up to that great height for the space of twenty minutes. After the display is over the water settles down in the basin several inches and the temperature slowly falls to 150°. We called this the Grand Geyser, for its power seemed greater than any other of which we obtained any knowledge in the valley. There are two orifices in one basin; one of them seems to have no raised rim, and is a very modest-looking spring in a state of quiescence, and no one would for a moment suspect the power that was temporarily slumbering below. The orifice is oblong, two and a half by four feet, while for the space of ten feet in every direction around it are rounded masses of silica, from a few inches to three feet in diameter, looking like spongiform corals. Nothing could exceed the crystal clearness of the water. This is the Grand Geyser. Within twenty feet of this orifice is a second one, of irregular quadrangular form, fifteen by twenty-five feet; the east side of the main outer rim of reservoir extended twenty feet beyond the large orifice. The bottom of this great reservoir is covered with thick spongiform masses, and in addition the rim is most elegantly adorned with countless pearl-like beads, of all sizes. There are several beautiful triangular reservoirs, one and one half by three feet, set around the outer sides of the rim, with numerous smaller ones, full of clear water, with hundreds of small depressions most beautifully scalloped. As we recede from the rim, the waters as they pass slowly away produce, by evaporation, broad shallow basins, with thin, elegantly colored partitions, portions of which have the form of toad-stools. When the water settles into these depressions, or flows away toward the river in numerous small channels, the wonderful variety of coloring which is so attractive to the eye is produced. The larger orifice seems to be in a state of violent agitation as often as once in twenty minutes, raising up the entire mass of water ten or fifteen feet. It is never altogether quiet. Although these two orifices are within the same rim, I could not ascertain that there is the slightest connection with each other. When the large orifice is much agitated it does not disturb the equanimity of the Grand Geyser. They both operate perfectly independently of each other. Indeed I do not know that there is a connection between any of the springs in the whole basin, though there may be in some rare cases. The Grand Geyser operated twice while we were in the basin, with an interval of about thirty-two hours; of course the displays could not be exactly periodic, but it would be an interesting study to remain several days to watch carefully the movements of such a power."

Just east of Grand Geyser is the Saw-mill, a moderate sized geyser, with three smaller ones by the side of it, all playing at the same time. From the larger a column of water is constantly shot up fifteen or twenty feet, with much the sound of the escape of steam from a pipe. The orifice is not more than six inches in diameter; but with the three smaller ones playing at the same time a great commotion is excited. Near this little group are several large boiling springs, which throw up the water in the centre from two to four feet. These are funnel-shaped, with orifices from six inches to two feet in diameter, in basins with nearly circular rims, from fifteen to forty feet in diameter. About one fourth of a mile northeast of the Castle, upon a mound thirty feet about above the river, built up of thin laminæ of silica, and rounded off, rise four chimneys of different sizes, which are geysers, though perhaps not spouting extensively at present. One is twelve inches high, nearly circular, and three feet in diameter; the second is oblong, four by six feet, with rather coarsely scolloped margins, with an aperture about fifteen inches in diameter; the third chimney is about three feet high and six feet in diameter at the base, with an orifice nearly quadrangular twelve inches across. The spongiform masses inside are covered all over with beautiful pearly beads of silica. The fourth chimney rises five feet above the mound, is ten feet in diameter at the base, with an orifice two feet across, lined inside with the spongiform masses. This has been at one time a first-class geyser, but is now fast going to decay, a beautiful ruin. The elegant bead work on the margin, and all the spongiform masses, now are falling into pieces, forming great quantities of débris around the base of the mound. There is also one boiling spring of great beauty. The orifice, which is nearly circular and beautifully scolloped around the margin, extends straight down, and the water rises within an inch or two of the margin. The water is in a state of constant agitation, boiling up two feet at times. The margin has a coating of bright cream-yellow, while all around the surface there is the most delicate and intricate embroidering, surpassing the most elaborate lace-work. Surrounding the crater is an outer reservoir four feet wide, with a white and reddish-yellow rim, while in the bottom of the reservoir is the variegated sediment which aids in giving such a wonderfully gay appearance to the spring. A stream of water flows from the spring to the river, and the channel is lined for fifty yards with the variegated sediment. Near this is another mound which rises, with laminated steps, about six feet. Dr. Hayden called it the Bathtub. It has much the shape and size of an ordinary bathing-tub, five feet by ten, beautifully scolloped around the inner margin with the spongiform or cauliflower masses of silica, the outer surface being adorned with the greatest profusion of pearly beads. The water is constantly boiling up two feet high, though but a small quantity flows from the spring. The entire valley is full of similar springs, many of these no doubt geysers whose periods of activity have never been observed. "We could not distinguish," writes Lieut. Doane, "the geysers from the other hot springs except by seeing them play, and doubtless there are many besides in the valley of great size, which we saw when quiet, and classed as boiling springs. They all vary in times, force, deposits, and color of water. The number of springs of all kinds in the valley is not less than fifteen hundred; and, with the exception of the Bluestone Springs, scarcely any two are exactly alike. Taken as an aggregate, the Firehole Basin surpasses all the other great wonders of the continent. It produces an effect on the mind of the beholder utterly staggering and overpowering. During the night we were several times awakened by the rush of steam and the hissing of the waters, as the restless geysers spouted forth in the darkness. A constant rumbling, as of machinery in labor, filled the air, which was damp and warm throughout the night."

Lieutenant Doane's suspicion that many quiet-looking springs were slumbering geysers, was speedily and grandly justified. The very next morning his company were awakened by a fearful hissing sound, accompanied by the rush of falling water. On looking out, they saw on the other side of the stream a small crater, three feet in height, with an opening 26 inches in diameter, which had scarcely been noticed on the previous day. Now it was playing a perpendicular jet to the height of 219 feet, amid great clouds of steam, and causing the ground to tremble as the heavy body of water fell with tremendous splashes upon the shelly strata below. Huge masses of rock were torn from their places and borne away into the river channel. It played thus, steadily, for ten minutes, giving time to obtain an accurate measurement by triangulation. This crater gave no notice of being a geyser, and its appearance and size, compared with others, were altogether insignificant. "We were more than ever convinced," adds Lieutenant Doane, "that continued observation would develop the knowledge of geysers in greater numbers, and perhaps of greater projectile force than any we had seen."

Crossing the river once more to the south side—that of Old Faithful and the Castle—we find another large group of springs, the chief of which is the Giant. This is remarkable rather for its immense flow of water than for any peculiarity of structure. It has a rugged crater, shaped like the base of a broken horn, twelve feet high. Its cavity or nozzle is seven feet in diameter.

The Giant Geyser.

During its quiescent state the boiling water can be seen in its chambers at a depth of forty feet, the action of the steam and water together producing a loud, rumbling sound. Near, and acting in concert with it, are half a dozen smaller craters from two to eight feet in height, constantly full of water, and boiling violently from two to six feet into the air. "This great geyser," says Lieutenant Doane, "played several times while we were in the valley, on one occasion throwing constantly for over three hours a stream of water seven feet in diameter, from 90 to 200 feet perpendicularly. While playing it doubled the size of the Firehole River."

At the base of the mountain further south is a remarkable geyser, discovered by Colonel Barlow, and called by him the Comet.

The crater of this geyser is very beautiful, though, being but slightly elevated above the general slope of the plateau, it might easily be overlooked, should it not happen to play during the visit of an examining party. There are three openings. One, a very small aperture, emits puffs of steam, similar to the exhaust-pipe of a steam engine. The large one in the centre, six feet across, boils violently during an eruption, but does not throw water to a great height. The third opening is the geyser proper. It is twelve by eighteen inches in diameter, somewhat narrowed as it descends, and is of great depth—smooth and straight. These cavities are all lined with delicate deposit, beautifully enamelled, in appearance as delicate as frost-work, but hard and strong, requiring the assistance of a hammer to detach fragments for specimens. Soon after Colonel Barlow entered the basin he witnessed a grand eruption of this geyser. He says:

"A roar was heard near the hill-side a hundred yards distant, and upon rushing out in that direction we saw a huge mass of steam issuing from a crater at the base of the hill, accompanied by a column of water rising to a height far exceeding that of any geyser yet seen. This grand fountain continued to play for several minutes, when, having subsided, I approached to obtain a closer view of the aperture whence had issued such a powerful stream. A sudden gush of steam drove me away, following which the water was again impelled upward and upward, far above the steam, until it seemed to have lost the controlling force of gravity. The roar was like the sound of a tornado, but there was no apparent effort—a steady stream, very graceful and perfectly vertical, except as a slight breeze may have waved it to and fro. Strong and smooth, it continued to ascend, like the stream from a powerful steam fire-engine. We were all lost in astonishment at the sudden and marvellous spectacle. I have no hesitancy in stating that this geyser played to the height of over two hundred feet. It commenced at five o'clock in the afternoon and continued twenty minutes.

"The enthusiasm of the party as they watched this wondrous display knew no bounds. Those who were usually loud and boisterous in the exhibition of their feelings, became subdued and simply gazed in silent awe;

while the more sober members seemed to lose their natural gravity and manifested their delight in shouts of rapture. For myself, I remember trying to obtain a view of the fountain from all points of the compass at once, and was brought to a realizing sense of the difficulties attending the execution of this desire, by discovering that I was waiting in the torrent of hot water which was now flooding the nearly level surface of the surrounding rock.

"After the grand column of water subsided, vast clouds of steam, were for some time ejected from the throat of the geyser, and also from a small rent close beside the main orifice.

"During the following day we watched this crater with increasing vigilance in the hope of witnessing another of its stupendous exhibitions. The photographer kept his camera levelled upon the spot all day, and careful arrangements for triangulating the height of the column were also made. But though numerous indications of another eruption were observed at intervals by the rising of the water with violent ebullitions, no explosion occurred till about ten o'clock at night, when the grand performance was repeated. The spectacle by moonlight was truly sublime, but less satisfactory than in the day, since it was more difficult to distinguish between the column of water and the masses of vapor escaping with it. The interval between its eruptions was approximately determined as about twenty-nine hours; we therefore reasoned that it would play again at three o'clock the following night, and at eight on the succeeding morning. But we were doomed to disappointment; the eruption persisted in taking-place before daylight on the three succeeding nights, thus preventing the observations we so much desired to make."

Two hundred yards below the Giant is the Grotto so named from the curious irregularity of its inclosing walls. It has an exceedingly intricate formation, with fantastic arches, pillars, and turreted sides, and discharges several times a day. Several of the first explorers crawled through the sinuous apertures when all was quiet. It seemed as harmless as curious then, but their opinion was changed slightly when they saw it an hour after throwing a column of water six feet in diameter to the height of sixty feet. Near it are several vents in which the water boils constantly to the height of six feet, large streams running down the banks into the river.

Around the point of a hill a few hundred yards south of the Grotto, and partially concealed by a grove of pines, is a white cone twenty-five feet high and a hundred feet in diameter at the base. It has evidently been a geyser of considerable importance, but it now merely sends forth puffs of steam from a small orifice at the top. Near it is a quiet hot-spring with a most elegantly scalloped rim. Back of this Pyramid is the Punch Bowl, and still further south, the Black Sand Geyser, neither of which has been specially described.

Returning to the river and crossing we find at the water's edge, nearly opposite the Grotto, the Riverside Geyser, and a short distance below, on the same side, is the Fan. The latter geyser has a double orifice, which discharges five radiating jets to the height of sixty feet, the falling drops and spray giving the appearance of a feather fan. The effect is very beautiful. Its eruptions are frequent, lasting usually from ten to thirty minutes. A vent connected with it, about forty feet distant, expels dense masses of vapor fifty or sixty feet high, accompanied by loud, sharp reports, during the time the geyser is in action. Lieutenant Doane describes the curious action of these vents as follows:

"First the steam would rush from the upper crater, roaring violently, then this would suddenly cease, to be followed by a fan-like jet of water rising from the lower crater to the height of over forty feet, playing for perhaps two minutes; then this would suddenly stop flowing, and the steam would rush forth again for a time. Occasionally the small crater threw a transverse stream, alternating with the others; and thus they played on for hours, after which all would subside to a gentle bubbling."

Along both banks of the river are small craters built up in every conceivable shape. Several streams pour out cascades from round holes in the rocky bank of the river, and all around are little geysers playing at intervals from six to forty feet.

A plateau opposite the Fan contains fifteen hot springs of various characteristics; some are of a deep blue color and have fantastic caverns distinctly visible below the surface of the water. The openings at the surface are often beautifully edged with delicately wrought rock fringes. One variety deposits a red or brown leathery substance, partially adhering to the sides and bottom of the cavern and waving to and fro like water plants. In size these springs vary from five to forty feet in diameter.

Fan Geyser.

Two hundred yards below the Fan are two lively geysers called the Sentinels. The one on the right bank of the stream is in constant agitation, its waters revolving horizontally with great violence, occasionally spouting upward to the height of twenty feet, with a lateral projection of fifty feet. Much steam is thrown off at each eruption. The crater of this geyser is three feet by ten. The companion Sentinel on the other side of the stream is smaller and less active. At this point the river-valley is narrow and the stream rapid, with a considerable fall. Forty or fifty comparatively unimportant geysers and boiling springs are scattered along the narrow valley to the junction of Iron Spring Creek, the lower limit of the Upper Geyser Basin.

Iron Spring Creek, a stream about half the size of Firehole River, takes its name from a group of springs on its banks, about a mile south of the Giant. Among the most noticeable of these is a group of eight beautiful springs enclosed in a single rim, one hundred and forty feet in length. The interior of the basin is lined with a rose-colored deposit. These springs are situated on the crest of an eminence incrusted with rocky deposits which encroach on the adjacent forest, whose dead and whitened trunks bear evidence of the deadly effect of the hot water flowing among them. On a considerable mound, at the junction of Iron Creek with the main stream, is a group of geysers that do not differ materially from those already described. The central member of the group is known as Soda Geyser.

The Bee-Hive

Chapter 13

LOWER GEYSER BASIN—FIREHOLE RIVER

Between the Upper and Lower Geyser Basin is a space of two or three miles entirely free from hot springs; yet the abundance of spring deposit over all the valley shows that the region was once the scene of great thermal activity; the bottom over which the river flows is paved with silica. Vegetation grows remarkably rank along the stream, and in the valley where the crust of silica does not prevent it, the perpetual warmth caused by the proximity to the springs being very favorable to the growth of plants. The forest grows close down to the margin of the river, making travel very difficult, and in one place the hills of trachyte almost close in the valley.

At the upper end of the basin—which comprises an area of about thirty square miles—are three large boiling springs, on the west margin of the river; nearly opposite are three more, and a short distance below, on the same side, four or five more. Anywhere else these springs would be accounted marvels; but they are so eclipsed by a group a few rods further down the stream that we can give them only a passing glance. This group includes some of the grandest hot-springs in the world. The most formidable is near the margin of the river. Dr. Hayden says:

"It seems to have broken out close by the river, and to have continually enlarged its orifice by the breaking down of its sides. It evidently commenced on the east side, and the continual wear of the underside of the crust on the west side has caused the margin to fall in, until an aperture at least 250 feet in diameter has been formed, with walls or sides twenty to thirty feet high, showing the laminæ of deposition perfectly. The water is intensely agitated all the time, boiling like a caldron, from which a vast column of steam is ever rising, filling the orifice. As the passing breeze sweeps it away for a moment, one looks down into this terrible seething pit with terror. All around the sides are large masses of the silicious crust that have fallen from the rim. An immense column of water flows out of this caldron into the river. As it pours over the marginal slope, it descends by numerous small channels, with a large number of smaller ones spreading over a broad surface, and the marvellous beauty of the strikingly vivid coloring far surpasses anything of the kind we have seen in this land of wondrous beauty—every possible shade of color, from the vivid scarlet to a bright rose, and every shade of yellow to delicate cream, mingled with vivid green from minute vegetation. Some of the channels were lined with a very fine, delicate yellow, silky material, which vibrates at every movement of the waters. Mr. Thomas Moran, the distinguished artist, obtained studies of these beautiful springs, and from his well-known reputation as a colorist, we look for a painting that will convey some conception to the mind of the exquisite variety of colors around this spring. There was one most beautiful funnel-shaped spring, twenty feet in diameter at the top, but tapering down, lined inside and outside with the most delicate decorations. Indeed, to one looking down into its clear depths, it seemed like a fairy palace. The same jelly-like substance or pulp to which I have before alluded, covers a large area with the various shades of light-red and green. The surface yields to the tread like a cushion. It is about two inches in thickness, and although seldom so tenacious as to hold together, yet it may be taken up in quite large masses, and when it becomes dry it is blown about by the wind like fragments of variegated lichens."

Near this magnificent spring is a hill of silica with a spring 150 feet in diameter on its summit. It is known as the Cauldron. The water boils up in the centre, and overflows with such uniformity on all sides as to form a succession of ornamental steps, from one to three inches in height, just as water would freeze in flowing down a gentle declivity. It has the same transparent clearness, the same brilliancy of coloring, as the spring above described, but the hot steam and the thinness of the rim prevented an approach near enough to observe its depth, or ascertain its temperature, except at one edge, where it was 180°. The average temperature of twenty of the springs of this group was 184°.

About a mile below this group, on the west side of the river, are four small lakes, with quiet surfaces and water as blue as the sky. One of them is nearly half a mile in length. Their water is cold at the present time, but their basins give indications of their having once been enormous hot springs.

A mile or so further down the river is a group of a hundred or more important springs besides a countless number of unimportant and dead springs, covering a space of nearly a square mile. Only a few of them can be specially noticed here. One, on the right bank of the river, is called the Conch Spring, from the resemblance of its basin to a shell, eight feet by ten.

A little below the Conch Spring, on the very margin of the river, is a fine geyser, which has built for itself a crater three feet high, with a shell a foot thick. The inside of the crater is about six feet in diameter. The water is in constant agitation; sometimes it will boil up so violently as to throw the entire mass up four feet, and then it will die down so as to boil like a caldron. The water is perfectly clear, and the overflow forms a stream six inches wide and two inches deep, passing down the sides of the crater and thence into the river along a most exquisitely decorated channel. The entire surface of the crater is covered with pearl-like beads, formed by the spray. A section of the crater shows it to have been built up very slowly, in thin laminæ. Another spring has a crater like a horn, about a foot in diameter at the top and six feet at the base. It is called the Horn Geyser. It is in constant ebullition, and has the same ornamentations as the one just described. A spring on a level with the river has an enormous square basin, thirty feet across, of unknown depth. It is called the Bath Spring. A little below is another basin of wonderful beauty, called the Cavern. The water issues from several apertures beneath the crust near the margin of the river. The basin itself is fifteen by twenty feet, and twenty feet deep. Nothing can exceed the transparent clearness of the water; the slightest object is reflected in its clear depths, and the bright blue tints are indescribable. Mud springs are also numerous and important in this group. As usual, they are of all sizes, from an inch or two to twenty or thirty feet in diameter, with contents varying from turbid water to stiff mud. They seldom have any visible outlet, but are in constant agitation, with a sound which varies with the consistency of the contents; several give off a suppressed thud as the gases burst their way through the stiff mortar. Sometimes the mortar is as white as snow; sometimes brown, or tinged with a variety of vivid colors. One mud spring, located in the woods near a small lake, northeast of the Cavern, has a basin thirty by forty feet, with sides fifteen feet high. It is in constant action, frequently hurling the mud outside of the rim. All around it are a number of little vents, which keep up a simmering noise. Some of these vents have built up little cones, from four to twelve inches high, many of them sealed at the top. On removing the cone, the inside is found to be lined with delicate crystals of sulphur, deposited from the steam.

On the opposite side of the river, along a little branch that flows in from the west, is a considerable group of geysers and boiling springs. Near the base of the mountains is a first-class boiling spring with a curious fungus-like rim. It is always in violent agitation, sending forth great columns of steam. It flows from beneath a hill, and is surrounded with springs whose silicious deposits take the form of the toad-stool fungus. Some of this group may be called spasmodic springs. One, with a most beautifully scalloped rim, fifteen by twenty feet in diameter, is always boiling, and occasionally explodes with great violence, shooting its water several feet into the air.

Along the eastern side of this Lower Geyser Basin are several extensive areas abounding in mud-springs, boiling springs and geysers, whose infinitely varied characteristics can have no more than the briefest notice. Beginning at the north, the first that commands attention is that whose central object of interest is the Thud Geyser, so called from the peculiar noise it makes as the water rises and recedes. It is situated on the slope of a small hill, is about twenty feet in diameter, and has a crater five feet wide and five high, composed of geyserite of a greyish color, full of deep pockets containing balls of the same material, about the size of walnuts, each one covered with little rosette-like formations. The column of water thrown out by this geyser during its eruptions is very wide, and reaches the height of fifty feet. Near it was obtained some pieces of wood, coated with geyserite of a delicate pink tinge: the silica had thoroughly penetrated the woody fibre. There were found, also, pine-cones coated in the same manner, forming beautiful specimens. A few yards back of the geyser are three large mud-springs, in one of which the mud is red, in another white, and in the third pink—the jets of steam causing the mud to assume the form of small conical points throughout the basins. They are situated in a bed of clay, the red color being due to iron. Below these latter are some chalybeate springs whose bright-red iron deposit have spread over a considerable area, in glaring contrast with the white color of the silicious deposits of their neighbors.

In the same group is a fissure spring forty feet long, four feet wide and ten deep, clear as crystal; also a large basin nearly circular, fifty feet in diameter with a number of huge apertures, some of which throw up water thirty feet. One orifice shoots a constant stream six feet high. All around this geyser-group are smaller springs continually bubbling, and large numbers of small geysers, some constantly playing to heights not exceeding ten feet, while others merely keep up a violent ebullition, rising and falling with a pulsating motion. There is also one beautiful quiet spring, with a basin so large that it looks like a small lake. Into its clear depths one may look down thirty or forty feet, beholding a fairy-like palace adorned with more brilliant colors than any structure made by human hands. The aggregate waters of the group form a little stream which flows westward into the small lake already noticed in connection with the mud-springs at the lower end of the basin.

South of the Thud Geyser is a large basin 150 feet in diameter, enclosing a crater twenty-five feet in diameter. From the inner crater the water is thrown up in a vast column sixty feet high, falling back in silver-white globules, a natural fountain of marvellous beauty. A short distance south of this Fountain Geyser is the most remarkable mud-pot in the Firehole valley. Its surface, forty by sixty feet in diameter, is covered with large puffs, which, in exploding with suppressed thuds, throw the hot mud several feet into the air and spatter the broad rounded rim in every direction. The mud is an impalpable silicious clay, of every shade of color from the purest white to a bright pink. Within a few feet of this mud-spring is a large clear spring sixty feet across, with perhaps fifty centres of ebullition. It is filled with the rusty, leathery deposit already described, and all around the basin where the waters overflow is an extensive deposit of iron.

A quarter of a mile east of the mud-springs, at the northwestern base of a mountain-spur and extending a thousand yards up a ravine, is a group of springs occupying a space five hundred yards wide. One of these, the Fissure Spring, is a hundred feet long and from four to ten feet wide. Quite a large stream flows from this spring. Many of the surrounding springs remain full to the rim, and are in constant ebullition, yet no water flows from them. Others discharge great quantities of water. In this group are three sulphur springs, the only ones in the region: the sulphur present however is not very abundant. Silica and iron seem to be the dominant constituents of nearly all the deposits. Some of the springs send forth a disagreeable odor, and deposit a curious black sediment like fine gunpowder. Near the centre of the group is a small lake 600 feet long and 150 wide. By its eastern shore is a geyser which spouts very regularly to the height of fifteen or twenty feet. West of the lake are two small geysers cones incrusted with a cauliflower-like formation; near them in a fissure are balls of geyserite coated in the same manner.

A thousand yards further south, in the southeastern corner of the basin, is a ravine a mile and a half long and three hundred yards wide, occupied by a most interesting group of springs and geysers. Just below the mouth of the ravine, on a mound fifteen feet high, is a large cone twenty-five feet high, probably a geyser. Steam issues steadily from the top with the sound of a high-pressure engine. It is called the White Dome Geyser.

In this lower basin there are very few raised craters, the most of the springs and geysers having funnel-shaped basins, with rims of various forms, but mostly circular. In this group there is besides the White Dome a small cone with its top broken off. It is four feet in diameter, with an aperture eighteen inches across. It is called the Bee-hive, for several feet around on all sides the surface of the ground is ornamented with pearly tubercles of silica, from the size of a pea to three inches in diameter. The spring basins in this group have every variety of form. One, a fine boiling spring, has an oval rim five feet by eight, its sides running straight down beyond the reach of vision. Another is funnel-shaped, tapering to a narrow aperture, with a scalloped rim, projecting several inches over the water. Some springs discharge no water; others send forth a stream two feet wide and six inches deep. In one of the streams, the channel of which is about two feet wide and one foot deep, the water is filled with a plant with a pinkish-yellow base, bordered with a very fine green silky fringe, perpetually vibrating with the flowing waters. "Except that they were a rich vegetable green, these fringes had the form and texture of the finest cashmere wool. The luxuriant growth of vegetation in and along the borders of these little streams," adds Dr. Hayden, "was a wonder of beauty. The whole view was there superior to anything of the kind I had seen."

In some of the springs Dr. Hayden's assistants found butterflies which had fallen in and been scalded to death. On taking them out they were found to be partially petrified, and coated with silica.

At the mouth of the ravine is the principal geyser of the group. Its basin is circular and about 60 feet in diameter, although the spring itself, which is in the centre, is only about 15 or 20 feet in diameter. The incrusted margin is full of sinuses, filled with hot water, which falls into them whenever the geyser is in operation. These pockets contain also smooth pebbles of geyserite, varying in size from that of a pea to a large-sized walnut, rounded by the action of the water. The water in the spring of the geyser is of a blue color and in constant agitation, though more violently so just before spouting. The column of water projected reaches the height of 100 feet, and is accompanied by immense clouds of steam.

Not far from this geyser is an elegantly scalloped spring, nearly circular, twenty-five feet in diameter, and with vertical sides to an unknown depth. The entire mass of the water is most violently agitated at times, and, overflowing the sides of the basin, passes off in terrace pools or reservoirs to the main stream, producing a system of architecture out of silica similar to that of the calcareous springs on Gardiner's River. The gay colors, from bright pink to delicate rose, are well shown.

The valley is filled with springs up to its very source; and springs which burst from the mountain side, eight hundred feet above the level of the basin, have temperatures ranging from 166° to 180°. Tracing one exceptionally cool stream up the south side of the cañon, Dr. Hayden found on the almost vertical side of the mountain a little spring so imbedded in bright green moss that it could hardly be seen.

"With great difficulty," he says, "we managed to climb up the mountain side, and, clearing away the moss, obtained the first water that we could drink for eight hours. In all of our examination during the day we had not found a drop of water of sufficiently low temperature to take into our mouths, though there were hundreds of the most beautiful springs all around us. We were like Coleridge's mariner in the great ocean, 'Water, water everywhere, but not a drop to drink.'"

Looking back over the valley the morning before his departure for the Upper Basin, Dr. Hayden saw it literally filled with columns of steam, ascending from more than a thousand vents. "I can compare the view," he writes, "to nothing but that of some manufacturing city like Pittsburgh, as seen from a high point, except that instead of the black coal smoke, there are here the white delicate clouds of steam. Small groups or solitary springs that are scattered everywhere in the woods, upon the mountain-sides, and which would otherwise escape observation, are detected by the columns of steam. It is evident that some of these groups of springs have changed their base of operations within a comparatively recent period; for about midway on the east side of the lower basin there is a large area covered with a thick, apparently modern, deposit of the silica, as white as snow, while standing quite thickly all around are dead pines, which appear to have been destroyed by the excessive overflow of water and the increased deposition. These dry trees have a most desolate look; many of them have fallen down and are incrusted with the silica, while portions that have fallen into the boiling springs have been reduced to pulp. This seems to be one of the conditions of silicification, for when these pulpy masses of wood are permitted to dry by the cessation of the springs, the most perfect specimens of petrified wood are the result. In one instance a green pine-tree had fallen so as to immerse its thick top in a large hot basin, and leaves, twigs, and cones had become completely incrusted with the white silica, and a portion had entered into the cellular structure, so that when removed from the water, and dried in the sun, very fair specimens were obtained. Members of my party obtained specimens of pine cones that were sufficiently silicified to be packed away among the collections."

Grasshoppers, and even butterflies, as we have seen, are occasionally subjected to the same treatment, with the same result. By-and-by, when the geyser regions become a popular resort, the preparation of petrifactions to be carried away as mementos, may become quite an item of entertainment if not of industry.

To obtain a complete view of the Lower Geyser Basin, Colonel Barlow made a trip to the summit of the Twin Buttes on the west side of the basin. From this point the valley of Firehole River could be overlooked for a distance of twenty miles; but nothing new was discovered except an attractive fall plunging over a precipice a short distance to the south.

After much severe climbing over rocky ridges, and scrambling through deep and thickly wooded ravines, he succeeded in reaching the foot of the fall—the loveliest vision he had ever beheld. "Towering above my head," he writes, "was a perpendicular cliff, three hundred feet high, while from a slight depression in its upper edge descended a sparkling stream, dashed into spray as it impinged against projecting angles of the rocky wall. On reaching the bottom the mist is gathered into a shallow basin, forming a pool of clear cold water, delightfully refreshing in this region of steaming geysers and volcanic heat. After resting a moment in this quiet retreat, the water slowly finds its way through the forest, and crossing the geyser valley eventually reaches Firehole River, some two miles distant."

From the marshy ground about the fall the pines shoot upward to an astonishing altitude, as though ambitious to overtop the cliff. Colonel Barlow approached the fall through a natural avenue in these pines, and as he caught sight of its dancing water, leaping with life-like action down the face of the overhanging cliff, the thought of fairies was so strongly suggested that he could think of no name so appropriate as the Fall of the Fairies.

The extreme north of the Lower Basin is bounded by the East Fork of Madison River, along whose valley, within the basin and above it, are numerous groups of interesting springs, though not materially different from those already described. Near the head of the stream sulphur springs are abundant, with here and there extensive deposits of sulphur. Steam-vents are frequent, their orifices lined with sulphur, and the surrounding crust filled with crystals of a vivid yellow. The channels of the streams are lined with a delicate veil of creamy sulphur. In some of the springs, lower down the stream, iron predominates. Within the basin on the south side of the East Fork are a hundred springs or more, any one of which, if alone, would be worthy of elaborate description. In some the fallen leaves of trees are frosted with silica as white as snow, and the inner surface of the basins are covered with a delicate bead-like embroidery of marvellous beauty. The most beautiful of the group is a Prismatic Spring, like those described in the Upper Basin. “Nothing ever conceived by human art,” says Dr. Hayden, “could equal the peculiar vividness and delicacy of coloring of these remarkable prismatic springs.” About a mile south of the East Fork, at the head of a little stream that flows into the Firehole, is another of these brilliant springs. A thin, delicately ornamental rim of silica surrounds a basin six feet in diameter, filled to the brim with water of marvellous transparency. When its surface is rippled by a passing breeze, the reflected sunlight is broken as by a million prisms. It is called the Rainbow Spring.

Chapter 14

NATURAL HISTORY OF GEYSERS AND OTHER THERMAL SPRINGS

In Icelandic speech the word *geyser* means simply *rager*, and is applied indiscriminately to all turbulent fountains of water or mud. The most violent and noisy "rager" on the island being the great intermittent spouting spring near Haukadal, it naturally gained for itself the title, The Geyser; and being the earliest known and most remarkable fountain of the kind, its native common name was adopted in other languages as the generic name for all springs of its class.

The history of this great, but no longer the greatest geyser, begins in the early part of the fifteenth century, when its eruptions are mentioned in Icelandic records. In the middle of the seventeenth century the Bishop of Skalholt noticed its daily discharges. A hundred years later Olafsen and Povelsen described it as having three or four eruptions a day, some of them attaining the height of 300 feet, including, doubtless, the uprush of vapor. The depth of its tube was then 72 feet; now it is commonly given as 74, though Commander Forbes, R. N., claims that it is not so deep by ten feet. In 1774 Von Troil estimated the height of the ejected column at 92 feet. Seventeen years later, Stanley gave 96 feet as its greatest height. Forty yards west of the Geyser this traveller found a rival, called the Strokr, (in English, the Churn,) playing to the height of 130 feet. The same year, 1789, an earthquake destroyed the mechanism of the Strokr, converting it into a quiet reservoir of boiling water, whereupon its name was transferred to the present Strokr, which then became especially active and noisy. In 1804 the Geyser had regained somewhat of its ancient power, erupting every six hours to the height of 200 feet; and the original Strokr had repaired its tube so that it could lift a column to nearly the same height and sustain it for a much longer period. During the next five years the power of the Geyser fell off a half, and its paroxysms became much less frequent—Hooker estimating its column, in 1809, at 100 feet, and Mackenzie, a year later, at 90 feet, its eruptions taking place every thirty hours. At the same time the Strokr played every ten or twelve hours, sixty feet high, for the space of thirty minutes. In 1815 the periods had changed again, the Geyser erupting every six hours, to an average height of 80 feet,—the jets occasionally reaching 150 feet, while the Strokr had prolonged its quiet intervals to twenty-four hours. Of late years the Geyser's violent eruptions seldom occur oftener than once in thirty hours, and do not exceed 100 feet in altitude, and generally averaging 70 or 80 feet. Between these eruptions are usually two minor spirts, attaining from 30 to 50 feet. The Strokr is exceedingly irregular in its operation, and generally requires a dose of turf to bring on an exhibition.

A grand eruption of the Geyser has been admirably described by Commander Forbes.

"Twice during the night," he says, "I was aroused by the unearthly complaints of *The* Geyser; but beyond the vast clouds of vapor which invariably follow each detonation, and a gentle overflowing of the basin, they were false alarms. As morning was breaking it sounded an unmistakable 'reveille,' which would have roused the dead: and I had barely time to take up my position at the brink of the old 'Strokr' before full power was turned on. Jet succeeded jet with fearful rapidity, earth trembled and the very cone itself seemed to stagger under the ordeal. Portions of its sides, rent with the uncontrollable fury it had suddenly generated, were ripped off and flew up in volleys, soaring high above water and steam, whilst the latter rolled away in fleecy clouds before the light north wind, and catching the rays of the morning sun just glistening over the Jökul tops in the East, was lustrous white as the purest snow. Discharge succeeded discharge in rapid succession for upwards of four minutes, when, apparently exhausted and its basin empty, I scrambled up to the margin, intending to have a good look down the tube, which I imagined must also be empty; but the water was still within a few feet of the brink, and boiling furiously. Hastening back to my former position, the basin filled rapidly, and I was just in time to witness the most magnificent explosion of all. Everything seemed to depend on this superhuman effort, and a solid, unbroken column of water twenty-five feet in circumference, was hurled upwards, attaining an altitude very near 100 feet. Here the column paused for a moment before reversing its motion, then fell listless and exhausted through the volumes which followed it into its throbbing cup, again to undergo its fiery ordeal at the threshold of the infernal regions."

Grand as this display must have been, it was but a momentary spasm, a feeble effort compared with the terrific force which sustains the Giant's river-volume, with a steady uprush two hundred feet high, for the space of three hours and a half. There are many, perhaps scores, of geysers in the Firehole Basin, which—even in midsummer, when their action is weakest—far surpass the glory of Iceland.

But what is the origin of the power that sustains these wonderful eruptions? And what is the cause of its intermittent action? Fortunately these questions are not only answerable, but the answers are susceptible of demonstration, as Professor Tyndall has shown in his admirable lectures on heat considered as a mode of motion, wherein he gives the following lucid description of the mechanism and development of the Great Geyser of Iceland: in principle the description applies equally to the geysers of Firehole Basin, and all other springs of the kind.

"It consists of a tube seventy-four feet deep and ten feet in diameter. The tube is surmounted by a basin which measures from north to south fifty-two feet across, and from east to west sixty feet. The interior of the tube and basin is coated with a beautiful smooth silicious plaster, so hard as to resist the blows of a hammer; and the first question is, how was this wonderful tube constructed—how was this perfect plaster laid on? Chemical analysis shows that the water holds silica in solution, and the conjecture might therefore arise that the water had deposited the silica against the sides of the tube and basin. But this is not the case: the water deposits no sediment; no matter how long it may be kept, no solid substance is separated from it. It may be bottled up and preserved for years as clear as crystal, without showing the slightest tendency to form a precipitate. To answer the question in this way would moreover assume that the shaft was formed by some foreign agency, and that the water merely lined it. The geyser basin, however, rests upon the summit of a mound about forty feet high, and it is evident from mere inspection that the mound has been deposited by the geyser. But in building up this mound the spring must have formed the tube which perforates the mound, and hence the conclusion that the geyser is the architect of its own tube.

If we place a quantity of geyser water in an evaporating basin the following takes place: in the centre of the basin the liquid deposits nothing, but at the sides where it is drawn up by capillary attraction, and thus subjected to speedy evaporation, we find silica deposited. Round the edge a ring of silica is laid on, and not until the evaporation has continued a considerable time do we find the slightest turbidity in the middle of the water. This experiment is the microscopic representative of what occurs in Iceland. Imagine the case of a simple thermal silicious spring, whose waters trickle down a gentle inclosure; the water thus exposed evaporates speedily, and silica is deposited. This deposit gradually elevates the side over which the water passes until finally the latter has to take another course. The same takes place here, the ground is elevated as before, and the spring has to move forward. Thus it is compelled to travel round and round, discharging its silica and deepening the shaft in which it dwells, until finally, in the course of ages, the simple spring has produced this wonderful apparatus which has so long puzzled and astonished both the traveller and the philosopher."

The time required for the construction of the Great Geyser tube has been estimated by Commander Forbes as ten or eleven centuries, on the following grounds: a bunch of grass, placed under a little fall made by the ejected water, receives, in twenty-four hours, a coating of silica the thickness of a thin sheet of paper, or about one five-hundredth part of an inch. At this rate it would take 1036 years to build up the 762 inches, which, according to his measurement, is the depth of the tube. In evidence of the probable truth of this estimate he makes note of the following facts: first, there is no notice of this fountain in the early history of the colonization of the island 986 years ago, at which time the tube would have been only three feet deep, and its eruptions too slight to attract attention; second, 436 years afterwards, when the tube would have been twenty-six feet deep, and the eruptions proportionately important, the Geyser is mentioned; third, accurate records of all occurrences were kept by the early inhabitants, and if so remarkable a phenomenon had existed at the time, it could not have been left unnoticed.

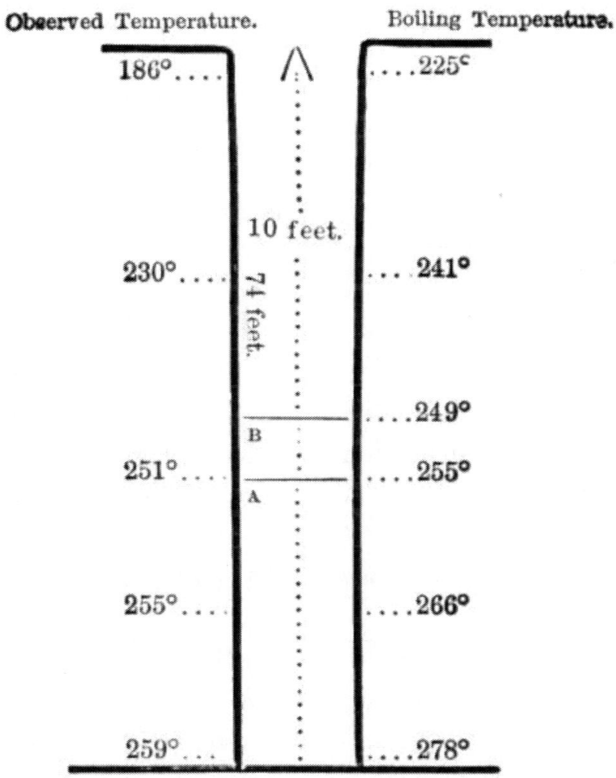

The phenomena attending a geyser-eruption—the filling of the basin with water, the agitation of the water, with deafening detonations, the escape of steam, and so on—have been sufficiently described in the preceding chapters. Their causes have been ingeniously explained by Professor

Bunsen, who succeeded in determining the temperature of the geyser-tube, throughout its entire length, a few minutes before an eruption. The annexed diagram shows on the left the observed temperatures of the water at different depths, and on the right the temperatures at which water would boil, taking into account the pressure of the atmosphere increased by the presence of the superincumbent column of water. The degrees have been changed from Centigrade to our familiar Fahrenheit standard, disregarding fractions.

It will be observed that in no part of the tube does the water reach the boiling point. The nearest approach to it is at A, thirty feet from the bottom; out even here the water is some four degrees below the temperature at which it could boil. How then is an eruption possible?

Professor Tyndall's explanation is in substance this: Suppose that by the entrance of steam from the ducts near the bottom of the tube the geyser column is elevated six feet, a height quite within the limits of actual observation; the water at A is thereby transferred to B. Its boiling point at A is 255°, and its actual temperature is 251°; but at B its boiling point is only 249°, hence when transferred from A to B, the heat which it possesses is in excess of that necessary to make it boil. This excess of heat is instantly applied to the generation of steam; the column is thus lifted higher, and the water below is relieved of pressure, and its boiling point lowered. More steam is generated; this lifts the column still higher, and compels the generation of more and more steam, until the whole upper portion of the column bursts into ebullition, and the water, mixed with steam-clouds, is projected into the atmosphere, and we have the geyser eruption in all its grandeur.

One confirmation of this theory of Bunsen's is that small stones suspended in the lower part of the geyser-tube are not thrown out during an eruption; and a stronger confirmation lies in the fact that all the peculiarities of geyser action can be imitated. Professor Tyndall uses for this purpose an apparatus consisting of a tube of iron six feet long, surmounted by a basin, and heated by fires underneath. To imitate, as far as possible, the conditions of the geyser, he encircles the tube with a second fire, two feet from the bottom. As the water in the tube becomes heated, the phenomena of geyser eruption are repeated in miniature with beautiful regularity. By stopping the mouth of the tube with a cork, the enforced explosions of the Strokr are reproduced; and by similar simple devices the action of all other eruptive springs may be accurately imitated.

All through the Firehole Basin and around Yellowstone Lake are many extinct geysers; sometimes, as in the case of Old Faithful, an active geyser is surrounded by a number of deserted cones, the remains of ancient "roarers." What occasions their decline? Earthquakes may, and no doubt frequently do, derange their mechanism, as observed in the old Strokr. But most of them probably die a natural death, from old age and decrepitude.

"A moment's reflection," says Professor Tyndall, "will suggest that there must be a limit to the operations of the geyser. When the tube has reached such an altitude that the water in the depths below, owing to the increased pressure, cannot attain its boiling point, the eruptions of necessity cease. The spring, however, continues to deposit its silica, and often forms a *Laug*, or cistern. Some of these, in Iceland, are forty feet deep. Their beauty, according to Bunsen, is indescribable. Over the surface curls a light vapor; the water is of the purest azure, and tints with its lovely hue the fantastic incrustations on the cistern walls; while at the bottom is often seen the mouth of the once mighty geyser. There are in Iceland vast, but now extinct geyser operations. Mounds are observed whose shafts are filled with rubbish, the water having forced a passage underneath and retired to other scenes of action. We have, in fact, the geyser in its youth, manhood, old age, and death here presented to us. In its youth as a simple thermal spring; in its manhood, as an eruptive column; in its old age, as the tranquil *Laug*; while its death is recorded by the ruined shaft and mound, which testify the fact of its once active existence."

All that Professor Tyndall describes so eloquently of Iceland, exists in our Grand National Park in infinitely greater variety, and magnitude, and splendor. And much more: Iceland has no Gardiner's River. To find the nearest approach to the marvels of White Mountain Hot Spring, we must go to the opposite side of the globe—to New Zealand. In the celebrated Lake District of the North Island is a region of hot springs, far exceeding in extent and variety all the others in the world, save those of the Yellowstone. First of all, says Hochstetter, the most marvellous of the Rotomahana marvels is the Te Tarata—the Tattooed Rock—with its terraced marble steps projecting into the lake. The spring lies about eighty feet above the lake, on a fern-clad hill-slope, in a crater-like excavation, with steep reddish sides, from thirty to forty feet high, and open only toward the lake. The basin of the spring is about eighty feet long and sixty wide, filled to the brim with perfectly transparent water, which in its snow-white basin appears of a beautiful blue, like the blue turquoise. Immense clouds of steam curl up from the surface, obstructing the view, but the noise of boiling and seething is always audible. The water is slightly salt, but not unpleasant to the taste, chemically neutral, and possesses petrifying, or rather incrusting qualities, in a high degree. The deposit is silicious, like that of the Iceland springs and the springs around Yellowstone Lake, not calcareous, like those of Gardiner's River; yet the system of terraces built up by the deposit on the hill-slope has the same appearance of a cataract plunging over a series of natural shelves and suddenly turned to stone. The deposits cover an area of about three acres, a mere trifle compared with the square miles of similar formations on Gardiner's River and in the Yellowstone Basin.

In the same neighborhood is a system of bubbling mud-pools, miniature copies of those on the Yellowstone above the falls. The principal group, lying in a ravine nearly a quarter of a mile long, is described by Dr. Hochstetter as follows:

"The entrance to the ravine is overgrown with a thicket and rather difficult of access; it also requires considerable caution, as suspicious places have to be passed, where the visitor is in danger of being swallowed up in heated mud. Inside, the ravine has the appearance of a volcanic crater. The bare walls, utterly destitute of vegetation, are terribly fissured and torn, and odd-looking rocky serratures, threatening every moment to break loose, loom up like dismal spectres from red, white and blue fumarole-clay—evidently the last remains of decomposed rocks. The bottom of the ravine is of fine mud, scattered with blocks of silicious deposit, like cakes of floating ice after a thaw. Here, a big caldron of mud is simmering; there, lies a deep basin of boiling water; next to this is a terrible hole, emitting hissing jets of steam, and further on are mud-cones from two to five feet high, vomiting hot mud from their craters with dull rumblings, and imitating on a small scale the play of large fire volcanoes." The gay colors of the Yellowstone mud-springs are frequently exhibited in the volcanic lake district of New Zealand, and so indeed are most of the other phenomena we have been studying, though on a far less magnificent scale. For example, the grandest "Firehole basin" on the island occupies the Shallow valley of a little stream the Waikato, for the distance of a mile. It is but a cabinet exhibition comparatively, yet the learned geologist of the *Novara* expedition grows eloquent in his description of it. "In the morning a dense fog lay upon the Waikato, but it soon vanished, the sun shone brightly into the valley, and now—what a sight! In its swift course, forming rapids after rapids, the Waikato was plunging through the deep valley between steep-rising mountains; its floods whirling and foaming round two small rocky islands in the middle of the river, were dashing with a loud uproar through the defile of the valley. Along its banks white clouds of steam were ascending from hot cascades falling into the river, and from basins full of boiling water shut in by a white mass of stone. Yonder a steaming fountain was rising and falling; now there sprung from another place a second fountain; this also ceased in its turn; then two commenced playing simultaneously, one quite low at the river bank, the other opposite on a terrace; and thus the play continued with endless changes, as though experiments were being made with grand waterworks, to see whether the fountains were all in perfect order, and whether the waterfalls had a sufficient supply. I began to count the places where a boiling waterbasin was visible, or where a cloud of steam indicated the existence of one. I counted seventy-six points, without, however, being able to survey the whole region, and among them were numerous intermittent geyser-like fountains with periodical eruptions of water."

The picture is admirably drawn, but could the artist have done so well with the stupendous chasm of the Grand Cañon? or the thousand volcanic vents of Firehole Basin with their deafening detonations, their immeasurable evolutions of water and steam? It is possible, but scarcely probable. The incomprehensible grandeur of the scene would have awed, astounded, bewildered him, and like our Yellowstone explorers, he would have despaired of grouping the myriad marvels into one grand effect, and contented himself with setting down a few details of form and color.

The Great Cañon and Lower Falls of the Yellowstone.

In following the exploration of the Yellowstone country and Firehole Basin, the reader has doubtless observed, in the passage from the quiet springs of Gardiner's River to the erupting fountains further on, that there is a complete change in the nature of the deposits. The mounds and terraces built up by the former have for their basis *lime*, those of the latter *silica*. Dr. Hayden attributes the calcareous deposits to the deep bed of limestone underlying the springs, but not all waters have the power of dissolving lime so freely, nor could ordinary water take from the trachytic lavas below the silicious springs around Yellowstone Lake and in the Firehole Basin, the silica that appears so abundantly in their deposits. There must be other forces at work. What are they? "Both kinds of springs," says Dr. Hochstetter, in his chapter on New Zealand springs, "owe their origin to the water permeating the surface and sinking through fissures into the bowels of the earth, where it becomes heated by the still existing volcanic fires. High-pressure steam is thus generated, which, accompanied by volcanic gases—such as muriatic acid, sulphurous acid, sulphuretted hydrogen and carbonic acid—rises again toward the colder surface and is there condensed into hot water. The overheated steam and the gases decompose the rock beneath, dissolving certain ingredients which are deposited on the surface. According to Bunsen's ingenious observations, a chronological succession takes place in the coöperation of the gases. The sulphurous acid acts first. It is generated where rising sulphur vapor comes in contact with glowing masses of rock. Wherever vapors of sulphurous acid are constantly formed, there acid-springs or solfataras arise. Incrustations of alum are very common in such places, arising from the action of sulphuric acid on the alumina and alkali of the lavas; another product of the decomposition of the lavas is gypsum, or sulphate of lime, the residuum being a more or less ferruginous fumarole clay, the material of the mud-pools. After the sulphurous acid comes sulphuretted hydrogen, produced by the action of steam on sulphids; and by the mutual decomposition of sulphuretted hydrogen and sulphurous acid sulphur is formed, the characteristic precipitate in all solfataras, while the deposition of silica is either entirely wanting or quite inconsiderable, and the smell of sulphuretted hydrogen is but rarely noticed. These acid springs have no periodical outbursts of water.

In course of time the source of sulphurous acid becomes exhausted, and sulphuretted hydrogen alone remains active. The acid reaction of the soil disappears, yielding to an alkaline reaction by the formation of sulphids. At the same time carbonic acid begins to act upon the rocks, and the alkaline bi-carbonates thus produced dissolve the silica, which on the evaporation of the water is deposited in the form of opal or quartz or silicious earth, and thus the shell of the springs is formed, on the structure of which the periodicity of the outburst depends.... The deposition of silica in quantities sufficient for the formation of this spring-apparatus in the course of years, takes place only in the alkaline springs. Their water is either neutral or has a slightly alkaline reaction. Silica, common salt, carbonates and sulphates are the chief ingredients dissolved in it. In the place of sulphurous acid the odor of sulphuretted hydrogen is sometimes observed in these springs.... By the gradual cooling of the volcanic rocks under the surface of the earth the hot springs themselves gradually die out, for they too are but a transient phenomenon in the eternal change of created things.”

Chapter 15

MR. EVERTS'S THIRTY-SEVEN DAYS OF PERIL

On the day that I found myself separated from the company, and for several days previous, our course had been impeded by the dense growth of pine forest, and occasional large tracts of fallen timber frequently rendering our progress almost impossible. Whenever we came to one of these immense windfalls, each man engaged in the pursuit of a passage through it, and it was while thus employed, and with the idea that I had found one, that I strayed out of sight and hearing of my comrades. We had had a toilsome day. It was quite late in the afternoon. As separations like these had frequently occurred, it gave me no alarm, and I rode on, fully confident of soon rejoining the company, or of finding their camp. I came up with the pack-horse, which Mr. Langford afterwards recovered, and tried to drive him along. But failing to do so, and my eyesight being defective, I spurred forward, intending to return with assistance from the party. This incident tended to accelerate my speed. I rode on in the direction which I supposed had been taken, until darkness overtook me in the dense forest. This was disagreeable enough, but caused me no alarm. I had no doubt of being with the party at breakfast the next morning. I selected a spot for comfortable repose, picketed my horse, built a fire, and went to sleep.

The next morning I rose at early dawn, saddled and mounted my horse, and took my course in the supposed direction of the camp. Our ride of the previous day had been up a peninsula jutting into the lake, for the shore of which I started, with the expectation of finding my friends camped on the beach. The forest was quite dark, and the trees so thick, that it was only by a slow process I could get through them at all. In searching for the trail I became somewhat confused. The falling foliage of the pines had obliterated every trace of travel. I was obliged frequently to dismount, and examine the ground for the faintest indications. Coming to an opening, from which I could see several vistas, I dismounted for the purpose of selecting one leading in the direction I had chosen, and leaving my horse unhitched, as had always been my custom, walked a few rods into the forest. While surveying the ground my horse took fright, and I turned around in time to see him disappearing at full speed among the trees. That was the last I ever saw of him. It was yet quite dark. My blankets, gun, pistols, fishing-tackle, matches—everything, except the clothing on my person, a couple of knives, and a small opera-glass were attached to the saddle.

I did not yet realize the possibility of a permanent separation from the company. Instead of following up the pursuit of their camp, I engaged in an effort to recover my horse. Half a day's search convinced me of its impracticability. I wrote and posted in an open space several notices, which, if my friends should chance to see, would inform them of my condition and the route I had taken, and then struck out into the forest in the supposed direction of their camp. As the day wore on without any discovery, alarm took the place of anxiety at the prospect of another night alone in the wilderness, and this time without food or fire. But even this dismal foreboding was cheered by the hope that I should soon rejoin my companions, who would laugh at my adventure, and incorporate it as a thrilling episode into the journal of our trip. The bright side of a misfortune, as I found by experience, even under the worst possible circumstances, always presents some features of encouragement. When I began to realize that my condition was one of actual peril, I banished from my mind all fear of an unfavorable result. Seating myself on a log, I recalled every foot of the way I had travelled since the separation from my friends, and the most probable opinion I could form of their whereabouts was, that they had, by a course but little different from mine, passed by the spot where I had posted the notices, learned of my disaster, and were waiting for me to rejoin them there, or searching for me in that vicinity. A night must be spent amid the prostrate trunks before my return could be accomplished. At no time during my period of exile did I experience so much mental suffering from the cravings of hunger as when, exhausted with this long day of fruitless search, I resigned myself to a couch of pine foliage in the pitchy darkness of a thicket of small trees. Naturally timid in the night, I fully realized the exposure of my condition. I peered upward through the darkness, but all was blackness and gloom. The wind sighed mournfully through the pines. The forest seemed alive with the screeching of night birds, the angry barking of coyotes, and the prolonged, dismal howl of the gray wolf. These sounds, familiar by their constant occurrence throughout the journey, were now full of terror, and drove slumber from my eye-lids, but above all this, however, was the hope that I should be restored to my comrades the next day.

Early the next morning I rose unrefreshed, and pursued my weary way over the prostrate trunks. It was noon when I reached the spot where my notices were posted. No one had been there. My disappointment was almost overwhelming. For the first time, I realized that I was lost. Then came a crushing sense of destitution. No food, no fire; no means to procure either; alone in an unexplored wilderness, one hundred and fifty miles from the nearest human abode, surrounded by wild beasts, and famishing with hunger. It was no time for despondency. A moment afterwards I felt how calamity can elevate the mind, in the formation of the resolution "not to perish in that wilderness."

The hope of finding the party still controlled my plans. I thought, by traversing the peninsula centrally, I would be enabled to strike the shore of the lake in advance of their camp, and near the point of departure for the Madison. Acting upon this impression, I rose from a sleepless couch, and pursued my way through the timber-entangled forest. A feeling of weakness took the place of hunger. Conscious of the need of food, I felt no cravings. Occasionally, while scrambling over logs and through thickets, a sense of faintness and exhaustion would come over me, but I would suppress it with the audible expression, "This won't do; I *must* find my company." Despondency would sometimes strive with resolution for the mastery of my thoughts. I would think of home—of my daughter—and of the possible chance of starvation, or death in some more terrible form; but as often as these gloomy forebodings came, I would strive to banish them with reflections better adapted to my immediate necessities. I recollect at this time discussing the question, whether there was not implanted by Providence in every man a principle of self-preservation equal to any emergency which did not destroy his reason. I decided this question affirmatively a thousand times afterwards in my wanderings, and I record this experience here, that any person who reads it, should he ever find himself in like circumstances, may not despair. There is life in the thought. It will revive hope, allay hunger, renew energy, encourage perseverance, and, as I have proved in my own case, bring a man out of difficulty, when nothing else can avail.

It was mid-day when I emerged from the forest into an open space at the foot of the peninsula. A broad lake of beautiful curvature, with magnificent surroundings, lay before me, glittering in the sunbeams. It was full twelve miles in circumference. A wide belt of sand formed the margin which I was approaching, directly opposite to which, rising seemingly from the very depths of the water, towered the loftiest peak of a range of mountains apparently interminable. The ascending vapor from innumerable hot springs, and the sparkling jet of a single geyser added the feature of novelty to one of the grandest landscapes I ever beheld. Nor was the life of the scene less noticeable than its other attractions. Large flocks of swans and other waterfowl were sporting on the quiet surface of the lake; otters in great numbers performed the most amusing aquatic evolutions; mink and beaver swam around unscared, in most grotesque confusion. Deer, elk, and mountain sheep stared at me, manifesting more surprise than fear at my presence among them. The adjacent forest was vocal with the songs of birds, chief of which were the chattering notes of a species of mocking-bird, whose imitative efforts afforded abundant merriment. Seen under favorable circumstances, this assemblage of grandeur, beauty, and novelty would have been transporting; but jaded with travel, famishing with hunger, and distressed with anxiety, I was in no humor for ecstasy. My tastes were subdued and chastened by the perils which environed me. I longed for food, friends, and protection. Associated with my thoughts, however, was the wish that some of my friends of peculiar tastes could enjoy this display of secluded magnificence, now, probably, for the first time beheld by mortal eyes.

The lake was at least one thousand feet lower than the highest point of the peninsula, and several hundred feet below the level of Yellowstone Lake. I recognized the mountain which overshadowed it as the landmark which, a few days before, had received from General Washburn the name of Mount Everts; and as it is associated with some of the most agreeable and terrible incidents of my exile, I feel that I have more than a mere discoverer's right to the perpetuity of that christening. The lake is fed by innumerable small streams from the mountains, and the countless hot springs surrounding it. A large river flows from it, through a cañon a thousand feet in height, in a southeasterly direction, to a distant range of mountains, which I conjectured to be Snake River; and with the belief that I had discovered the source of the great southern tributary of the Columbia, I gave it the name of Bessie Lake, after the "Sole daughter of my house and heart."

During the first two days, the fear of meeting with Indians gave me considerable anxiety; but, when conscious of being lost, there was nothing I so much desired as to fall in with a lodge of Bannocks or Crows. Having nothing to tempt their cupidity, they would do me no personal harm, and, with the promise of reward, would probably minister to my wants and aid my deliverance. Imagine my delight, while gazing upon the animated expanse of water, at seeing sail out from a distant point a large canoe containing a single oarsman. It was rapidly approaching the shore where I was seated. With hurried steps I paced the beach to meet it, all my energies stimulated by the assurance it gave of food, safety, and restoration to friends.

As I drew near to it it turned towards the shore, and oh! Bitter disappointment, the object which my eager fancy had transformed into an angel of relief stalked from the water, an enormous pelican, flapped its dragon wings as if in mockery of my sorrow, and flew to a solitary point farther up the lake. This little incident quite unmanned me. The transition from joy to grief brought with it a terrible consciousness of the horrors of my condition. But night was fast approaching, and darkness would come with it. While looking for a spot where I might repose in safety, my attention was attracted to a small green plant of so lively a hue as to form a striking contrast with the deep pine foliage. For closer examination I pulled it up by the root, which was long and tapering, not unlike a radish. It was a thistle. I tasted it; it was palatable and nutritious. My appetite craved it, and the first meal in four days was made on thistle-roots. Eureka! I had found food. No optical illusion deceived me this time; I could subsist until I rejoined my companions. Glorious counterpoise to the wretchedness of the preceding half-hour!

Overjoyed at this discovery, with hunger allayed, I stretched myself under a tree, upon the foliage which had partially filled a space between contiguous trunks, and fell asleep. How long I slept I know not; but suddenly I was roused by a loud, shrill scream, like that of a human being in distress, poured, seemingly, into the very portals of my ear. There was no mistaking that fearful voice. I had been deceived by and answered it a dozen times while threading the forest, with the belief that it was a friendly signal. It was the screech of a mountain lion, so alarmingly near as to cause every nerve to thrill with terror. To yell in return, seize with convulsive grasp the limbs of the friendly tree, and swing myself into it, was the work of a moment. Scrambling hurriedly from limb to limb, I was soon as near the top as safety would permit. The savage beast was snuffing and growling below, apparently on the very spot I had just abandoned. I answered every growl with a responsive scream. Terrified at the delay and pawing of the beast, I increased my voice to its utmost volume, broke branches from the limbs, and, in the impotency of fright, madly hurled them at the spot whence the continued howlings proceeded.

Failing to alarm the animal, which now began to make the circuit of the tree, as if to select a spot for springing into it, I shook, with a strength increased by terror, the slender trunk until every limb rustled with the motion. All in vain. The terrible creature pursued his walk around the tree, lashing the ground with his tail, and prolonging his howlings almost to a roar. It was too dark to see, but the movements of the lion kept me apprised of its position. Whenever I heard it on one side of the tree I speedily changed to the opposite—an exercise which, in my weakened state, I could only have performed under the impulse of terror. I would alternately sweat and thrill with horror at the thought of being torn to pieces and devoured by this formidable monster. All my attempts to frighten it seemed unavailing. Disheartened at its persistency, and expecting every moment it would take the deadly leap, I tried to collect my thoughts, and prepare for the fatal encounter which I knew must result. Just at this moment it occurred to me that I would try silence. Clasping the trunk of the tree with both arms, I sat perfectly still. The lion, at this time ranging round, occasionally snuffing and pausing, and all the while filling the forest with the echo of his howlings, suddenly imitated my example. This silence was more terrible, if possible, than the clatter and crash of his movements through the brushwood, for now I did not know from what direction to expect his attack. Moments passed with me like hours. After a lapse of time which I cannot estimate, the beast gave a spring into the thicket and ran screaming into the forest. My deliverance was effected.

Had strength permitted, I should have retained my perch till daylight, but with the consciousness of escape from the jaws of the ferocious brute came a sense of overpowering weakness which almost palsied me, and made my descent from the tree both difficult and dangerous. Incredible as it may seem, I lay down in my old bed, and was soon lost in a slumber so profound that I did not awake until after daylight. The experience of the night seemed like a terrible dream; but the broken limbs which in the agony of consternation I had thrown from the tree, and the rifts made in the fallen foliage by my visitant in his circumambulations, were too convincing evidences of its reality. I could not dwell upon my exposure and escape without shuddering and reflecting that probably like perils would often occur under less fortunate circumstances, and with a more fatal issue. I wondered what fate was in reserve for me—whether I would ultimately sink from exhaustion and perish of starvation, or become the prey of some of the ferocious animals that roamed these vast fastnesses. My thoughts then turned to the loved ones at home. They could never know my fate, and would indulge a thousand conjectures concerning it, not the least distressing of which would be that I had been captured by a band of hostile Sioux, and tortured to death at the stake.

I was roused from this train of reflections by a marked change in the atmosphere. One of those dreary storms of mingled snow and rain, common to these high latitudes, set in. My clothing, which had been much torn, exposed my person to its "pitiless peltings." An easterly wind, rising to a gale, admonished me that it would be furious and of long duration. None of the discouragements I had met with dissipated the hope of rejoicing my friends; but foreseeing the delay, now unavoidable, I knew that my escape from the wilderness must be accomplished, if at all, by my own unaided exertions. This thought was terribly afflicting, and brought before me, in vivid array, all the dreadful realities of my condition. I could see no ray of hope. In this condition of mind I could find no better shelter than the spreading branches of a spruce tree, under which, covered with earth and boughs, I lay during the two succeeding days; the storm, meanwhile, raging with unabated violence. While thus exposed, and suffering from cold and hunger, a little benumbed bird, not larger than a snow-bird, hopped within my reach. I instantly seized and killed it, and, plucking its feathers, ate it raw. It was a delicious meal for a half-starved man.

Taking advantage of a lull in the elements, on the morning of the third day I rose early and started in the direction of a large group of hot springs which were steaming under the shadow of Mount Everts. The distance I travelled could not have been less than ten miles. Long before I reached the wonderful cluster of natural caldrons, the storm had recommenced. Chilled through, with my clothing thoroughly saturated, I lay down under a tree upon the heated incrustation until completely warmed. My heels and the sides of my feet were frozen. As soon as warmth had permeated my system, and I had quieted my appetite with a few thistle-roots, I took a survey of my surroundings, and selected a spot between two springs sufficiently asunder to afford heat at my head and feet. On this spot I built a bower of pine branches, spread its incrusted surface with fallen foliage and small boughs, and stowed myself away to await the close of the storm. Thistles were abundant, and I had fed upon them long enough to realize that they would, for a while at least, sustain life. In convenient proximity to my abode was a small, round, boiling spring, which I called my dinner-pot, and in which, from time to time, I cooked my roots.

This establishment, the best I could improvise with the means at hand, I occupied seven days—the first three of which were darkened by one of the most furious storms I ever saw. The vapor which supplied me with warmth saturated my clothing with its condensations. I was enveloped in a perpetual steam bath. At first this was barely preferable to the storm, but I soon became accustomed to it, and before I left, though thoroughly parboiled, actually enjoyed it.

I had little else to do during my imprisonment but cook, think, and sleep. Of the variety and strangeness of my reflections it is impossible to give the faintest conception. Much of my time was given to devising means for escape. I recollected to have read, at the time of their publication, the narratives of Lieutenant Strain and Doctor Kane, and derived courage and hope from the reflection that they struggled with and survived perils not unlike those which environed me. The chilling thought would then occur, that they were not alone. They had companions in suffering and sympathy. Each could bear his share of the burden of misery which it fell to my lot to bear alone, and make it lighter from the encouragement of mutual counsel and aid in a cause of common suffering. Selfish as the thought may seem, there was nothing I so much desired as a companion in misfortune. How greatly it would alleviate my distress! What a relief it would be to compare my wretchedness with that of a brother sufferer, and with him devise expedients for every exigency as it occurred! I confess to the weakness, if it be one, of having squandered much pity upon myself during the time I had little else to do.

Nothing gave me more concern than the want of fire. I recalled everything I had ever read or heard of the means by which fire could be produced; but none of them were within my reach. An escape without it was simply impossible. It was indispensable as a protection against night attacks from wild beasts. Exposure to another storm like the one just over would destroy my life, as this one would have done, but for the warmth derived from the springs. As I lay in my bower anxiously awaiting the disappearance of the snow, which had fallen to the depth of a foot or more, and impressed with the belief that for want of fire I should be obliged to remain among the springs, it occurred to me that I would erect some sort of monument, which might, at some future day, inform a casual visitor of the circumstances under which I had perished. A gleam of sunshine lit up the bosom of the lake, and with it the thought flashed upon my mind that I could, with a lens from my opera-glasses, get fire from Heaven. Oh, happy, life-renewing thought! Instantly subjecting it to the test of experiment, when I saw the smoke curl from the bit of dry wood in my fingers, I felt, if the whole world were offered me for it, I would cast it all aside before parting with that little spark. I was now the happy possessor of food and fire. These would carry me through. All thoughts of failure were instantly abandoned. Though the food was barely adequate to my necessities—a fact too painfully attested by my attenuated body—I had forgotten the cravings of hunger, and had the means of producing fire. I said to myself, "I will not despair."

My stay at the springs was prolonged several days by an accident that befell me on the third night after my arrival there. An unlucky movement while asleep broke the crust on which I reposed, and the hot steam, pouring upon my hip, scalded it severely before I could escape. This new affliction, added to my frost-bitten feet, already festering, was the cause of frequent delay and unceasing pain through all my wanderings. After obtaining fire, I set to work making preparations for as early departure as my condition would permit. I had lost both knives since parting from the company, but I now made a convenient substitute by sharpening the tongue of a buckle which I cut from my vest. With this I cut the legs and counters from my boots, making of them a passable pair of slippers, which I fastened to my feet as firmly as I could with strips of bark. With the ravellings of a linen handkerchief, aided by the magic buckle-tongue, I mended my clothing. Of the same material I made a fish-line, which, on finding a piece of red tape in one of my pockets better suited to the purpose, I abandoned as a "bad job." I made of a pin that I found in my coat a fish-hook, and, by sewing up the bottoms of my boot-legs, constructed a very good pair of pouches to carry my food in, fastening them to my belt by the straps.

Thus accountered, on the morning of the eighth day after my arrival at the springs I bade them a final farewell, and started on my course directly across that portion of the neck of the peninsula between me and the southeast arm of Yellowstone Lake. It was a beautiful morning. The sun shone bright and warm, and there was a freshness in the atmosphere truly exhilarating. As I wandered musingly along, the consciousness of being alone, and having surrendered all hope of finding my friends, returned upon me with crushing power. I felt, too, that those friends, by the necessities of their condition, had been compelled to abandon all efforts for my recovery. The thought was full of bitterness and sorrow. I tried to realize what their conjectures were concerning my disappearance; but could derive no consolation from the long and dismal train of circumstances they suggested. Weakened by a long fast, and the unsatisfying nature of the only food I could procure, I know that from this time onward to the day of my rescue, my mind, though unimpaired in those perceptions needful to self-preservation, was in a condition to receive impressions akin to insanity. I was constantly travelling in dream-land, and indulging in strange reveries such as I had never before known. I seemed to possess a sort of duality of being, which, while constantly reminding me of the necessities of my condition, fed my imagination with vagaries of the most extravagant character. Nevertheless I was perfectly conscious of the tendency of these morbid influences, and often tried to shake them off, but they would ever return with increased force, and I finally reasoned myself into the belief that their indulgence, as it afforded me pleasure, could work no harm while it did not interfere with my plans for deliverance. Thus I lived in a world of ideal happiness, and in a world of positive suffering at the same time.

A change in the wind and an overcast sky, accompanied by cold, brought with them a need of warmth. I drew out my lens and touchwood, but alas! There was no sun. I sat down on a log to await his friendly appearance. Hours passed; he did not come. Night, cold freezing night, set in, and found me exposed to all its terrors. A bleak hill-side sparsely covered with pines afforded poor accommodations for a half-clad, famishing man. I could only keep from freezing by the most active exertion in walking, rubbing, and striking my benumbed feet and hands against the logs. It seemed the longest, most terrible night of my life, and glad was I when the approaching dawn enabled me to commence retracing my steps to Bessie Lake. I arrived there at noon, built my first fire on the beach, and remained by it, recuperating for the succeeding two days.

The faint hope that my friends might be delayed by their search for me until I could rejoin them now forsook me altogether. I made my arrangements independent of it. Either of three directions I might take would affect my escape, if life and strength held out. I drew upon the sand of the beach a map of these several courses with reference to my starting-point from the lake, and considered well the difficulties each would present. All were sufficiently defined to avoid mistake. One was to follow Snake River a distance of one hundred miles or more to Eagle Rock bridge; another, to cross the country between the southern shore of Yellowstone Lake and the Madison Mountains, by scaling which I could easily reach the settlements in the Madison Valley; and the other, to retrace my journey over the long and discouraging route by which I had entered the country. Of these routes the last mentioned seemed the least inviting, probably because I had so recently traversed it, and was familiar with its difficulties. I had heard and read so much concerning the desolation and elemental upheavals and violent waters of the upper valley of the Snake, that I dared not attempt to return in that direction. The route by the Madison Range, encumbered by the single obstruction of the mountain barrier, was much the shortest, and so most unwisely, as will hereafter appear, I adopted it.

Filling my pouches with thistle-roots, I took a parting survey of the little solitude that had afforded me food and fire the preceding ten days, and with something of that melancholy feeling experienced by one who leaves his home to grapple with untried adventures, started for the nearest point on Yellowstone Lake. All that day I travelled over timber-heaps, amid tree tops, and through thickets. At noon I took the precaution to obtain fire. With a brand which I kept alive by frequent blowing, and constant waving to and fro, at a late hour in the afternoon, faint and exhausted, I kindled a fire for the night on the only vacant spot I could find amid a dense wilderness of pines. The deep gloom of the forest, in the spectral light which revealed on all sides of me a compact and unending growth of trunks and an impervious canopy of sombre foliage; the shrieking of night-birds; the super naturally human scream of the mountain lion, the prolonged howl of the wolf, made me insensible to all other forms of suffering.

The burn on my hip was so inflamed that I could only sleep in a sitting posture. Seated with my back against a tree, the smoke from the fire almost enveloping me in its suffocating folds, I vainly tried, amid the din and uproar of this horrible serenade, to woo the drowsy god. My imagination was instinct with terror. At one moment it seemed as if, in the density of a thicket, I could see the blazing eyes of a formidable forest monster fixed upon me, preparatory to a deadly leap; at another I fancied that I heard the swift approach of a pack of yelping wolves through the distant brushwood, which in a few moments would tear me limb from limb. Whenever, by fatigue and weakness, my terrors yielded to drowsiness, the least noise roused me to a sense of the hideousness of my condition. Once, in a fitful slumber, I fell forward into the fire, and inflicted a wretched burn on my hand. Oh! With what agony I longed for day!

A bright and glorious morning succeeded the dismal night, and brought with it the conviction that I had been the victim of uncontrollable nervous excitement. I resolved henceforth to banish it altogether, and, in much better spirits than I anticipated, resumed my journey towards the lake. Another day of unceasing toil among the tree-tops and thickets overtook me, near sunset, standing upon a lofty headland jutting into the lake, and commanding a magnificent prospect of the mountains and valley over an immense area. In front of me, at a distance of fifty miles away, in the clear blue of the horizon, rose the arrowy peaks of the Three Tetons. On the right, and apparently in close proximity to the eminence I occupied, rolled the picturesque range of the Madison, scarred with clefts, ravines, gorges, and cañons, each of which glittered in the sunlight or deepened in shadow as the fitful rays of the descending luminary glanced along their varied rocky irregularities. Above where I stood were the lofty domes of Mounts Langford and Doane, marking the limits of that wonderful barrier which had so long defied human power in its efforts to subdue it. Rising seemingly from the promontory which favored my vision was the familiar summit of Mount Everts, at the base of which I had dwelt so long, and which still seemed to hold me within its friendly shadow. All the vast country within this grand enclosure of mountains and lake, scarred and seamed with the grotesque ridges, rocky escarpments, undulating hillocks, and miniature lakes, and steaming with hot springs, produced by the volcanic forces of a former era, lay spread out before me like a vast panorama.

I doubt if distress and suffering can ever entirely obliterate all sense of natural grandeur and magnificence. Lost in the wonder and admiration inspired by this vast world of beauties, I nearly forgot to improve the few moments of remaining sunshine to obtain fire. With a lighted brand in my hand, I effected a most difficult and arduous descent of the abrupt and stony headland to the beach of the lake. The sand was soft and yielding. I kindled a fire, and removing the stiffened slippers from my feet, attached them to my belt, and wandered barefoot along the sandy shore to gather wood for the night. The dry, warm sand was most grateful to my lacerated and festering feet, and for a long time after my wood-pile was supplied, I sat with them uncovered. At length, conscious of the need of every possible protection from the freezing night atmosphere, I sought my belt for the slippers, and one was missing. In gathering the wood it had become detached and was lost. Darkness was closing over the landscape, when, sorely disheartened with the thought of passing the night with one foot exposed to a freezing temperature, I commenced a search for the missing slipper. I knew I could not travel a day without it. Fearful that it had dropped into the lake, and been carried by some recurrent wave beyond recovery, my search for an hour among fallen trees and bushes, up the hill-side and along the beach, in darkness and with flaming brands, at one moment crawling on hands and feet into a brush-heap, another peering among logs and bushes and stones, was filled with anxiety and dismay. Success at length rewarded my perseverance, and no language can describe the joy with which I drew the cause of so much distress from beneath the limb that, as I passed, had torn it from my belt. With a feeling of great relief I now sat down in the sand, my back to a log, and listened to the dash and roar of the waves. It was a wild lullaby, but had no terrors for a worn-out man. I never passed a night of more refreshing sleep. When I awoke my fire was extinguished save a few embers, which I soon fanned into a cheerful flame. I ate breakfast with some relish, and started along the beach in pursuit of a camp, believing that if successful I should find directions what to do, and food to sustain me. The search which I was making lay in the direction of my pre-arranged route to the Madison Mountains, which I intended to approach at their lowest point of altitude.

Buoyed by the hope of finding food and counsel, and another night of undisturbed repose in the sand, I resumed my journey along the shore, and at noon found the camp last occupied by my friends on the lake. I struck their trail in the sand some time before I came to it. A thorough search for food in the ground and trees revealed nothing, and no notice to apprise me of their movements could be seen. A dinner-fork, which afterwards proved to be of infinite service in digging roots, and a yeast-powder can, which would hold half a pint, and which I converted into a drinking-cup and dinner-pot, were the only evidences that the spot had ever been visited by civilized man. "Oh!" thought I, "why did they forget to leave me food!" It never occurred to me that they might have cached it, as I have since learned they did, in several spots nearer the place of my separation from them. I left the camp in deep dejection, with the purpose of following the trail of the party to the Madison. Carefully inspecting the faint traces left of their course of travel, I became satisfied that from some cause they had made a retrograde movement from this camp, and departed from the lake at a point farther downstream. Taking this as an indication that there were obstructions above, I commenced retracing my steps along the beach. An hour of sunshine in the afternoon enabled me to procure fire, which, in the usual manner, I carried to my camping-place. There I built a fire, and to protect myself from the wind, which was blowing violently, lashing the lake into foam, I made a bower of pine boughs, crept under it, and very soon fell asleep. How long I slept I know not, but I was aroused by the snapping and cracking of the burning foliage, to find my shelter and the adjacent forest in a broad sheet of flame. My left hand was badly burned, and my hair singed closer than a barber would have trimmed it, while making my escape from the semicircle of burning trees. Among the disasters of this fire, there was none I felt more seriously than the loss of my buckle-tongue knife, my pin fish-hook, and tape fish-line.

The grandeur of the burning forest surpasses description. An immense sheet of flame, following to their tops the lofty trees of an almost impenetrable pine forest, leaping madly from top to top, and sending thousands of forked tongues a hundred feet or more athwart the midnight darkness, lighting up with lurid gloom and glare the surrounding scenery of lake and mountains, fills the beholder with mingled feelings of awe and astonishment. I never before saw anything so terribly beautiful. It was marvellous to witness the flash-like rapidity with which the flames would mount the loftiest trees. The roaring, cracking, crashing, and snapping of falling limbs and burning foliage was deafening. On, on, on travelled the destructive element, until it seemed as if the whole forest was enveloped in flame. Afar up the wood-crowned hill, the overtopping trees shot forth pinnacles and walls and streamers of arrowy fire. The entire hill-side was an ocean of glowing and surging fiery billows. Favored by the gale, the conflagration spread with lightning swiftness over an illimitable extent of country, filling the atmosphere with driving clouds of suffocating fume, and leaving a broad and blackened trail of spectral trunks shorn of limbs and foliage, smoking and burning, to mark the immense sweep of its devastation.

Resolved to search for a trail no longer, when daylight came I selected for a landmark the lowest notch in the Madison Range. Carefully surveying the jagged and broken surface over which I must travel to reach it, I left the lake and pushed into the midst of its intricacies. All the day, until nearly sunset, I struggled over rugged hills, through windfalls, thickets, and matted forests, with the rock-ribbed beacon constantly in view. As I advanced it receded, as if in mockery of my toil. Night overtook me with my journey half accomplished. The precaution of obtaining fire gave me warmth and sleep, and long before daylight I was on my way. The hope of finding an easy pass into the valley of the Madison inspired me with fresh courage and determination; but long before I arrived at the base of the range, I scanned hopelessly its insurmountable difficulties. It presented to my eager vision an endless succession of inaccessible peaks and precipices, rising thousands of feet sheer and bare above the plain. No friendly gorge or gully or cañon invited such an effort as I could make to scale this rocky barrier. Oh for the faith that could remove mountains! How soon should this colossal fabric open at my approach! What a feeling of helpless despair came over me with the conviction that the journey of the last two days had been in vain! I seated myself on a rock upon the summit of a commanding hill, and cast my eyes along the only route which now seemed tenable—down the Yellowstone. How many dreary miles of forest and mountain filled the terrible panorama! I thought that before accepting this discouraging alternative I would spend a day in search for a pass. Twenty miles at most would take me into the Madison Valley, and thirty more restore me to friends who had abundance. Supposing that I should find plenty of thistles, I had left the lake with a small supply, and that was entirely spent. I looked in vain for them where I then was.

While I was thus considering whether to remain and search for a passage or return to the Yellowstone, I experienced one of those strange hallucinations which many of my friends have misnamed insanity, but which to me was Providence. An old clerical friend, for whose character and counsel I had always cherished peculiar regard, in some unaccountable manner seemed to be standing before me, charged with advice which would relieve my perplexity. I seemed to hear him say, as if in a voice and with the manner of authority:

"Go back immediately, as rapidly as your strength will permit. There is no food here, and the idea of scaling these rocks is madness."

"Doctor," I rejoined, "the distance is too great. I cannot live to travel it."

"Say not so. Your life depends upon the effort. Return at once. Start now, lest your resolution falter. Travel as fast and as far as possible—it is your only chance."

"Doctor, I am rejoiced to meet you in this hour of distress, but doubt the wisdom of your counsel. I am within seventy miles of Virginia. Just over these rocks, a few miles away, I shall find friends. My shoes are nearly worn out, my clothes are in tatters, and my strength is almost overcome. As a last trial, it seems to me I can but attempt to scale this mountain or perish in the effort, if God so wills."

"Don't think of it. Your power of endurance will carry you through. I will accompany you. Put your trust in Heaven. Help yourself and God will help you."

Overcome by these and other persuasions, and delighted with the idea of having a travelling companion, I plodded my way over the route I had come, intending at a certain point to change it so as to strike the river at the foot of the lake. Stopping after a few miles of travel, I had no difficulty in procuring fire, and passed a comfortable night. When I resumed my journey the next day the sun was just rising. Whenever I was disposed, as was often the case, to question the wisdom of the change of routes, my old friend appeared to be near with words of encouragement, but his reticence on other subjects both surprised and annoyed me. I was impressed at times, during the entire journey, with the belief that my return was a fatal error, and if my deliverance had failed should have perished with that conviction. Early this day I deflected from my old route and took my course for the foot of the lake, with the hope, by constant travel, to reach it the next day. The distance was greater than I anticipated. Nothing is more deceptive than distance in these high latitudes. At the close of each of the two succeeding days, my point of destination was seemingly as far from me as at the moment I took leave of the Madison Range, and when, cold and hungry, on the afternoon of the fourth day, I gathered the first food I had eaten in nearly five days, and lay down by my fire near the debouchure of the river, I had nearly abandoned all hope of escape.

At daybreak I was on the trail down the river. The thought I had adopted from the first, "I will cot perish in this wilderness," often revived my sinking spirits, when, from faintness and exhaustion, I felt but little desire for life. Once, while struggling through a field of tangled trunks which seemed interminable, at one of the pauses I found myself seriously considering whether it was not preferable to die there than renew the effort to proceed. I felt that all attempt to escape was but a bitter prolongation of the agony of dissolution. A seeming whisper in the air, "While there is life there is hope; take courage," broke the delusion, and I clambered on. I did not forget to improve the mid-day sun to procure fire. Sparks from the lighted brands had burned my hands and crisped the nails of my fingers, and the smoke from them had tanned my face to the complexion of an Indian. While passing through an opening in the forest I found the tip of a gull's wing; it was fresh. I made a fire upon the spot, mashed the bones with a stone, and consigning them to my camp kettle, the yeast-powder box, made half a pint of delicious broth. The remainder of that day and the night ensuing were given to sleep.

I lost all sense of time. Days and nights came and went, and were numbered only by the growing consciousness that I was gradually starving. I felt no hunger, did not eat to appease appetite, but to renew strength. I experienced but little pain. The gaping sores on my feet, the severe burn on my hip, the festering crevices at the joints of my fingers, all terrible in appearance, had ceased to give me the least concern. The roots which supplied my food had suspended the digestive power of the stomach, and their fibres were packed in it in a matted, compact mass.

Not so with my hours of slumber. They were visited by the most luxurious dreams. I would apparently visit the most gorgeously decorated restaurants of New York and Washington; sit down to immense tables spread with the most appetizing viands; partake of the richest oyster stews and plumpest pies; engage myself in the labor and preparation of curious dishes, and with them fill range upon range of elegantly furnished tables until they fairly groaned beneath the accumulated dainties prepared by my own hands. Frequently the entire night would seem to have been spent in getting up a sumptuous dinner. I would realize the fatigue of roasting, boiling, baking, and fabricating the choicest dishes known to the modern *cuisine*, and in my disturbed slumbers would enjoy with epicurean relish the food thus furnished even to repletion. Alas! There was more luxury than life in these somnolent vagaries.

It was a cold, gloomy day when I arrived in the vicinity of the falls. The sky was overcast and the snow-capped peaks rose chilly and bleak through the biting atmosphere. The moaning of the wind through the pines, mingling with the sullen roar of the falls, was strangely in unison with my own saddened feelings. I had no heart to gaze upon a scene which a few weeks before had inspired me with rapture and awe. One moment of sunshine was of more value to me than all the marvels amid which I was famishing. But the sun had hid his face and denied me all hope of obtaining fire. The only alternative was to seek shelter in a thicket. I penetrated the forest a long distance before finding one that suited me. Breaking and crowding my way into its very midst, I cleared a spot large enough to recline upon, interlaced the surrounding brushwood, gathered the fallen foliage into a bed, and lay down with a prayer for sleep and forgetfulness. Alas! Neither came. The coldness increased through the night. Constant friction with my hands and unceasing beating with my legs and feet saved me from freezing. It was the most terrible night of my journey, and when, with the early dawn, I pulled myself into a standing posture, it was to realize that my right arm was partially paralyzed, and my limbs so stiffened with cold as to be almost immovable. Fearing lest paralysis should suddenly seize upon the entire system, I literally dragged myself through the forest to the river. Seated near the verge of the great cañon below the falls, I anxiously awaited the appearance of the sun. That great luminary never looked so beautiful as when, a few moments afterwards, he emerged from the clouds and exposed his glowing beams to the concentrating powers of my lens. I kindled a mighty flame, fed it with every dry stick and broken tree-top I could find, and without motion, and almost without sense, remained beside it several hours. The great falls of the Yellowstone were roaring within three hundred yards, and the awful cañon yawned almost at my feet; but they had lost all charm for me. In fact, I regarded them as enemies which had lured me to destruction, and felt a sullen satisfaction in morbid indifference.

My old friend and adviser, whose presence I had felt more than seen the last few days, now forsook me altogether. But I was not alone. By some process which I was too weak to solve, my arms, legs, and stomach were transformed into so many travelling companions. Often for hours I would plod along conversing with these imaginary friends. Each had his peculiar wants which he expected me to supply. The stomach was importunate in his demand for a change of diet—complained incessantly of the roots I fed him, their present effect and more remote consequences. I would try to silence him with promises, beg of him to wait a few days, and when this failed of the quiet I desired, I would seek to intimidate him by declaring, as a sure result of negligence, our inability to reach home alive. All to no purpose—he tormented me with his fretful humors through the entire journey. The others would generally concur with him in these fancied altercations. The legs implored me for rest, and the arms complained that I gave them too much to do. Troublesome as they were, it was a pleasure to realize their presence. I worked for them, too, with right good will, doing many things for their seeming comfort which, had I felt myself alone, would have remained undone. They appeared to be perfectly helpless of themselves; would do nothing for me or for each other. I often wondered, while they ate and slept so much, that they did not aid in gathering wood and kindling fires. As a counterpoise to their own inertia, whenever they discovered languor in me on necessary occasions, they were not wanting in words of encouragement and cheer. I recall as I write, an instance where, by prompt and timely interposition, the representative of the stomach saved me from a death of dreadful agony. One day I came to a small stream issuing from a spring of mild temperature on the hill-side, swarming with minnows. I caught some with my hands, and ate them raw. To my taste they were delicious. But the stomach refused them, accused me of attempting to poison him, and would not be reconciled until I had emptied my pouch of the few fish I had put there for future use. Those that I ate made me very sick. Poisoned by the mineral in the water, had I glutted my appetite with them as

I intended, I should doubtless have died in the wilderness, in excruciating torment.

A gradual mental introversion grew upon me as physical weakness increased. The grand and massive scenery which, on the upward journey, had aroused every enthusiastic impulse of my nature, was now tame and spiritless. My thoughts were turned in upon myself—upon the dreadful fate which apparently lay just before me—and the possible happiness of the existence beyond.

Imaginary Companions.

All doubt of immortality fled in the light of present realities. So vivid were my conceptions of the future that at times I longed for death, not less as the beginning of happiness than as a release from misery. Led on by these reflections, I would recall the varied incidents of my journey—my escape from the lion, from fire, my return from the Madison Range—and in all of them I saw how much I had been indebted to that mysterious protection which comes only from the throne of the Eternal. And yet, starving, foot-sore, half blind, worn to a skeleton, was it surprising that I lacked the faith needful to buoy me above the dark waters of despair, which I now felt were closing around me?

In less serious moods, as I struggled along, my thoughts would revert to the single being on whom my holiest affections centred—my daughter. What a tie was that to bind me to life! Oh! Could I be restored to her for a single hour, long enough for parting counsel and blessing, it would be joy unspeakable! Long hours of painful travel were relieved of physical suffering by this absorbing agony of the mind, which, when from my present standpoint I contrast it with the personal calamities of my exile, swells into mountains.

To return from this digression. At many of the streams on my route I spent hours in endeavoring to catch trout, with a hook fashioned from the rim of my broken spectacles, but in no instance with success. The tackle was defective. The country was full of game in great variety. I saw large herds of deer, elk, antelope, occasionally a bear, and many smaller animals. Numerous flocks of ducks, geese, swans, and pelicans inhabited the lakes and rivers. But with no means of killing them, their presence was a perpetual aggravation. At all the camps of our company I stopped and recalled many pleasant incidents associated with them.

One afternoon, when approaching "Tower Falls," I came upon a large hollow tree, which, from the numerous tracks surrounding it, and the matted foliage in the cavity, I recognized as the den of a bear. It was a most inviting couch. Gathering a needful supply of wood and brush, I lighted a circle of piles around the tree, crawled into the nest, and passed a night of unbroken slumber. I rose the next morning to find that during the night the fires had communicated with the adjacent forest, and burned a large space in all directions, doubtless intimidating the rightful proprietor of the nest, and saving me from another midnight adventure.

At "Tower Falls" I spent the first half of a day in capturing a grasshopper, and the remainder in a fruitless effort to catch a mess of trout. In the agony of disappointment, I resolved to fish no more. A spirit of rebellion seized me. I determined that thistles should thenceforth be my only sustenance. "Why is it," I asked of myself, "that in the midst of abundance, every hour meeting with objects which would restore strength and vigor and energy, every moment contriving some device to procure the nourishment my wasting frame required, I should meet with these repeated and discouraging failures?" Thoughts of the early teaching of a pious mother suppressed these feelings. Oh! How often have the recollections of a loved New England home, and the memories of a happy childhood, cheered my sinking spirits, and dissipated the gathering gloom of despair! There were thoughts and feelings and mental anguishes without number, that visited me during my period of trial, that never can be known to any but my God and myself. Bitter as was my experience, it was not unrelieved by some of the most precious moments I have ever known.

Soon after leaving "Tower Falls," I entered the open country. Pine forests and windfalls were changed for sage brush and desolation, with occasional tracts of stunted verdure, barren hillsides, exhibiting here and there an isolated clamp of dwarf trees, and ravines filled with the rocky *debris* of adjacent mountains. My first camp on this part of the route, for the convenience of getting wood, was made near the summit of a range of towering foot-hills. Towards morning a storm of wind and snow nearly extinguished my fire. I became very cold; the storm was still raging when I arose, and the ground white with snow. I was perfectly bewildered, and had lost my course of travel. No visible object, seen through the almost blinding storm, reassured me, and there was no alternative but to find the river and take my direction from its current. Fortunately, after a few hours of stumbling and scrambling among rocks and over crests, I came to the precipitous side of the cañon through which it ran, and with much labor, both of hands and feet, descended it to the margin. I drank copiously of its pure waters, and sat beside it for a long time, waiting for the storm to abate, so that I could procure fire. The day wore on, without any prospect of a termination to the storm. Chilled through, my tattered clothing saturated, I saw before me a night of horrors unless I returned to my fire. The scramble up the side of the rocky cañon, in many places nearly perpendicular, was the hardest work of my journey. Often, while clinging to the jutting rocks with hands and feet, to reach a shelving projection, my grasp would unclose and I would slide many feet down the sharp declivity. It was night when, sore from the bruises I had received, I reached my fire; the storm, still raging, had nearly extinguished it. I found a few embers in the ashes, and with much difficulty kindled a flame. Here, on this bleak mountain side, as well as I now remember, I must have passed two nights beside the fire, in the storm. Many times during each night I crawled to the little clump of trees to gather wood, and brush, and the broken limbs of fallen tree-tops. All the sleep I obtained was snatched from the intervals which divided these labors. It was so harassed with frightful dreams as to afford little rest. I remember, before I left this camp, stripping up my sleeves to look at my shrunken arms. Flesh and blood had apparently left them. The skin clung to the bones like wet parchment. A child's hand could have clasped them from wrist to shoulder. "Yet," thought I, "it is death to remain; I cannot perish in this wilderness."

Taking counsel of this early formed resolution, I hobbled on my course through the snow, which was rapidly disappearing before the rays of the warm sun. Well knowing that I should find no thistles in the open country, I had filled my pouches with them before leaving the forest. My supply was running low, and there were yet several days of heavy mountain travel between me and Botelers' Ranch. With the most careful economy, it could last but two or three days longer. I saw the necessity of placing myself and imaginary companions upon allowance. The conflict which ensued with the stomach, when I announced this resolution, required great firmness to carry through. I tried wheedling and coaxing and promising; failing in these, I threatened to part company with a comrade so unreasonable, and he made no further complaint.

Two or three days before I was found, while ascending a steep hill, I fell from exhaustion into the sage brush, without the power to rise. Unbuckling my belt, as was my custom, I soon fell asleep. I have no idea of the time I slept, but upon awaking I fastened my belt, scrambled to my feet, and pursued my journey. As night drew on I selected a camping-place, gathered wood into a heap, and felt for my lens to procure fire. It was gone. If the earth had yawned to swallow me I would not have been more terrified. The only chance for life was lost. The last hope had fled. I seemed to feel the grim messenger who had been so long pursuing me knocking at the portals of my heart as I lay down by the side of the wood-pile, and covered myself with limbs and sage brush, with the dreadful conviction that my struggle for life was over, and that I should rise no more. The floodgates of misery seemed now to be opened, and it rushed in a living tide upon my soul. With the rapidity of lightning, I ran over every event of my life. Thoughts doubled and trebled upon me, until I saw, as if in vision, the entire past of my existence. It was all before me, as if painted with a sunbeam, and all seemingly faded like the phantoms of a vivid dream.

As calmness returned, reason resumed her empire. Fortunately, the weather was comfortable. I summoned all the powers of my memory, thought over every foot of the day's travel, and concluded that the glass must have become detached from my belt while sleeping. Five long miles over the hills must be retraced to regain it. There was no alternative, and before daylight I had staggered over half the distance. I found the lens on the spot where I had slept. No incident of my journey brought with it more of joy and relief.

Returning to the camp of the previous night, I lighted the pile I had prepared, and lay down for a night to rest. It was very cold, and towards morning commenced snowing. With difficulty I kept the fire alive. Sleep was impossible. When daylight came, I was impressed with the idea that I must go on despite the storm. A flash—momentary but vivid—came over me, that I should be saved. Snatching a lighted brand, I started through the storm. In the afternoon the storm abated and the sun shone at intervals. Coming to a small clump of trees, I set to work to prepare a camp. I laid the brand down which I had preserved with so much care, to pick up a few dry sticks with which to feed it, until I could collect wood for a camp-fire, and in the few minutes thus employed it expired. I sought to revive it, but every spark was gone. Clouds obscured the sun, now near the horizon, and the prospect of another night of exposure without fire became fearfully imminent. I sat down with my lens and the last remaining piece of touchwood I possessed to catch a gleam of sunshine, feeling that my life depended on it. In a few moments the cloud passed, and with trembling hands I presented the little disk to the face of the glowing luminary. Quivering with excitement lest a sudden cloud should interpose, a moment passed before I could hold the lens steadily enough to concentrate a burning focus. At length it came. The little thread of smoke curled gracefully up wards from the Heaven-lighted spark, which, a few moments afterwards, diffused with warmth and comfort my desolate lodgings.

I resumed my journey the next morning, with the belief that I should make no more fires with my lens. I must save a brand, or perish. The day was raw and gusty; an east wind, charged with storm, penetrated my nerves with irritating keenness. After walking a few miles the storm came on, and a coldness unlike any other I had ever felt seized me. It entered all my bones. I attempted to build a fire, but could not make it burn. Seizing a brand, I stumbled blindly on, stopping within the shadow of every rock and clump to renew energy for a final conflict for life. A solemn conviction that death was near, that at each pause I made my limbs would refuse further service, and that I should sink helpless and dying in my path, overwhelmed me with terror. Amid all this tumult of the mind, I felt that I had done all that man could do. I knew that in two or three days more I could affect my deliverance, and I derived no little satisfaction from the thought that, as I was now in the broad trail, my remains would be found, and my friends relieved of doubt as to my fate. Once only the thought flashed across my mind that I should be saved, and I seemed to hear a whispered command to "struggle on." Groping along the side of a hill, I became suddenly sensible of a sharp reflection, as of burnished steel. Looking up, through half-closed eyes, two rough but kindly faces met my gaze.

"Are you Mr. Everts?"

"Yes. All that is left of him."

"We have come for you."

"Who sent you?"

"Judge Lawrence and other friends."

"God bless him, and them, and you! I am saved!" and with these words, powerless of further effort, I fell forward into the arms of my preservers, in a state of unconsciousness. I was saved. On the very brink of the river which divides the known from the unknown, strong arms snatched me from the final plunge, and kind ministrations wooed me back to life.

Baronet and Prichette, my two preservers, by the usual appliances, soon restored me to consciousness, made a camp upon the spot, and while one went to Fort Ellis, a distance of seventy miles, to return with remedies to restore digestion and an ambulance to convey me to that post, the other sat by my side, and with all the care, sympathy, and solicitude of a brother, ministered to my frequent necessities. In two days I was sufficiently recovered in strength to be moved twenty miles down the trail to the cabin of some miners who were prospecting in that vicinity. From these men I received every possible attention which their humane and generous natures could devise. A good bed was provided, game was killed to make broth, and the best stores of their larder placed at my command. For four days, at a time when every day's labor was invaluable in their pursuit, they abandoned their work to aid in my restoration. Owing to the protracted inaction of the system, and the long period which must transpire before Prichette's return with remedies, my friends had serious doubts of my recovery.

The night after my arrival at the cabin, while suffering the most excruciating agony, and thinking that I had only been saved to die among friends, a loud knock was heard at the cabin door. An old man in mountain costume entered—a hunter, whose life was spent among the mountains. He was on his way to find a brother. He listened to the story of my sufferings, and tears rapidly coursed each other down his rough, weather-beaten face. But when he was told of my present necessity, brightening in a moment, he exclaimed:

"Why, Lord bless you, if that is all, I have the very remedy you need. In two hours' time all shall be well with you."

He left the cabin, returning in a moment with a sack filled with the fat of a bear which he had killed a few hours before. From this he rendered out a pint measure of oil. I drank the whole of it. It proved to be the needed remedy, and the next day, freed from pain, with appetite and digestion reestablished, I felt that good food and plenty of it were only necessary for an early recovery.

In a day or two I took leave of my kind friends, with a feeling of regret at parting, and of gratitude for their kindness as enduring as life.

Meeting the carriage on my way, I preceded to Boseman, where I remained among old friends, who gave me every attention until my health was sufficiently restored to allow me to return to my home at Helena.

My heartfelt thanks are due to the members of the expedition, all of whom devoted seven, and some of them twelve days to the search for me before they left Yellowstone Lake; and to Judge Lawrence, of Helena, and the friends who co-operated with him in the offer of reward which sent Baronet and Prichette to my rescue.

My narrative is finished. In the course of events the time is not far distant when the wonders of the Yellowstone will be made accessible to all lovers of sublimity, grandeur, and novelty in natural scenery, and its majestic waters become the abode of civilization and refinement; and when that arrives, I hope, in happier mood and under more auspicious circumstances, to revisit scenes fraught for me with such thrilling interest; to ramble along the glowing beach of Bessie Lake; to sit down amid the hot springs under the shadow of Mount Everts; to thread unscared the mazy forests, retrace the dreary journey to the Madison Range, and with enraptured fancy gaze upon the mingled glories and terrors of the great falls and marvellous cañon, and to enjoy, in happy contrast with the trials they recall, their power to delight, elevate, and overwhelm the mind with wondrous and majestic beauty.

Chapter 16

THE YELLOWSTONE PARK

As soon as Dr. Hayden could make known officially the results of his exploration of the Yellowstone Basin, action was begun to secure the reservation of a portion at least of the marvellous scenes which it embraces, for the undivided benefit, enjoyment and instruction of the country at large.

A bill to this effect was introduced into the Senate of the United States, on the 18th of December, 1871, by Hon. S. C. Pomeroy of Kansas. About the same time a similar bill was offered in the House of Representatives by Hon. William H. Claggett, delegate from Montana. The bill was referred to the Committees on Public Lands in both houses, who after due consideration returned with approbation the following report prepared by Dr. Hayden:

"The bill now before Congress has for its object the withdrawal from settlement, occupancy, or sale, under the laws of the United States, a tract of land fifty-five by sixty-five miles, about the sources of the Yellowstone and Missouri Rivers, and dedicates and sets it apart as a great national park or pleasure-ground for the benefit and enjoyment of the people. The entire area comprises within the limits of the reservation contemplated in this bill is not susceptible of cultivation with any degree of certainty, and the winters would be too severe for stock-raising. Whenever the altitude of the mountain districts exceeds 6,000 feet above tide-water, their settlement becomes problematical unless there are valuable mines to attract people. The entire area within the limits of the proposed reservation is over 6,000 feet in altitude, and the Yellowstone Lake, which occupies an area fifteen by twenty-two miles, or three hundred and thirty square miles, is 7,427 feet. The ranges of mountains that hem the valleys in on every side rise to the height of 10,000 and 12,000 feet and are covered with snow all the year. These mountains are all of volcanic origin, and it is not probable that any mines or minerals of value will ever be found there. During the months of June, July, and August the climate is pure and most invigorating, with scarcely any rain or storms of any kind, but the thermometer frequently sinks as low as 26°. There is frost every month of the year. This whole region was, in comparatively modern geological times, the scene of the most wonderful volcanic activity of any portion of our country. The hot springs and the geysers represent the last stages—the vents or escape-pipes—of these remarkable volcanic manifestations of the internal forces. All these springs are adorned with decorations more beautiful than human art ever conceived, and which have required thousands of years for the cunning hand of nature to form. Persons are now waiting for the spring to open to enter in and take possession of these remarkable curiosities, to make merchandise of these beautiful specimens, to fence in these rare wonders, so as to charge visitors a fee, as is now done at Niagara Falls, for the sight of that which ought to be as free as the air or water.

In a few years this region will be a place of resort for all classes of people from all portions of the world. The geysers of Iceland, which have been objects of interest for the scientific men and travellers of the entire world, sink into insignificance in comparison with the hot springs of the Yellowstone and Firehole Basins. As a place of resort for invalids, it will not be excelled by any portion of the world. If this bill fails to become a law this session, the vandals who are now waiting to enter into this wonder-land will, in a single season, despoil, beyond recovery, these remarkable curiosities, which have required all the cunning skill of nature thousands of years to prepare."

The Giantess.

"We have already shown that no portion of this tract can ever be made available for agricultural or mining purposes. Even if the altitude and the climate would permit the country to be made available, not over fifty square miles of the entire area could ever be settled. The valleys are all narrow, hemmed in by high volcanic mountains like gigantic walls.

The withdrawal of this tract, therefore, from sale or settlement takes nothing from the value of the public domain, and is no pecuniary loss to the Government, but will be regarded by the entire civilized world as a step of progress and an honor to Congress and the nation."

In the Senate the bill was ably advocated by Messrs. Pomeroy, Edmunds, Trumbull, Anthony and others. In the House the favorable remarks of Hon. H. L. Dawes were so clear and forcible that the bill was passed without opposition.

The text of the Act is as follows:

"*Be it enacted by the Senate and House of Representatives of the United States of America in Congress assembled*, That the tract of land in the Territories of Montana and Wyoming, lying near the headwaters of the Yellowstone River, and described as follows, to wit, commencing at the junction of Gardiner's River with the Yellowstone River, and running east to the meridian passing ten miles to the eastward of the most eastern point of Yellowstone Lake; thence south along said meridian to the parallel of latitude passing ten miles south of the most southern point of Yellowstone Lake; thence west along said parallel to the meridian passing fifteen miles west of the most western point of Madison Lake; thence north along said meridian to the latitude of the junction of the Yellowstone and Gardiner's River; thence east to the place of beginning, is hereby reserved and withdrawn from settlement, occupancy, or sale under the laws of the United States, and dedicated and set apart as a public park or pleasuring-ground for the benefit and enjoyment of the people; and all persons who shall locate or settle upon or occupy the same, or any part thereof, except as hereinafter provided, shall be considered trespassers and removed therefrom.

"Sec. 2. That said public park shall be under the exclusive control of the Secretary of the Interior, whose duty it shall be, as soon as practicable, to make and publish such rules and regulations as he may deem necessary or proper for the care and management of the same. Such regulations shall provide for the preservation, from injury or spoliation, of all timber, mineral deposits, natural curiosities or wonders within said park, and their retention in their natural condition. The Secretary may, in his discretion, grant leases for building purposes for terms not exceeding ten years, of small parcels of ground, at such places in said park as shall require the erection of buildings for the accommodation of visitors; all of the proceeds of said leases, and all other revenues that may be derived from any source connected with said park, to be expended under his direction in the management of the same, and the construction of roads and bridlepaths therein.

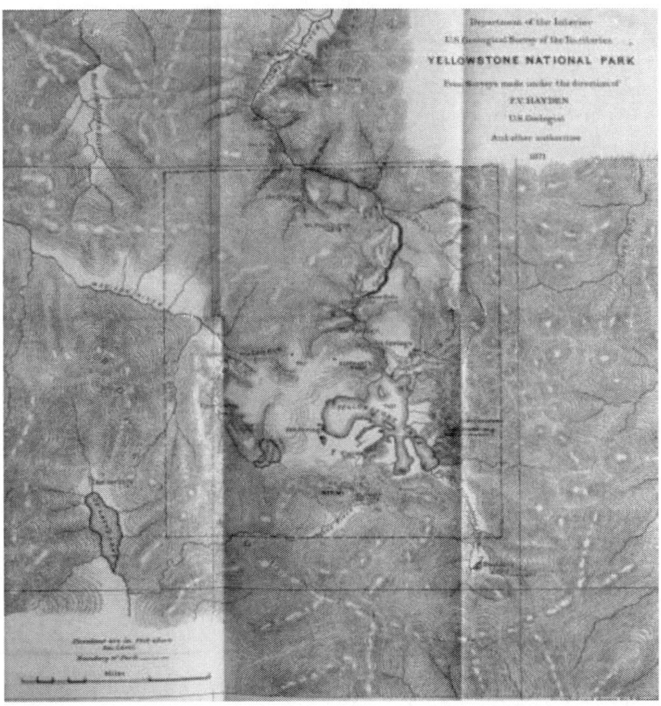

He shall provide against the wanton destruction of the fish and game found within said, park, and against their capture or destruction for the purposes of merchandise or profit. He shall also cause all persons trespassing upon the same after the passage of this act to be removed therefrom, and generally shall be authorized to take all such measures as shall be necessary or proper to fully carry out the objects and purposes of this act."

This Act was approved March 1, 1872; and shortly after the Hon. N. P. Langford, whose graphic descriptions of the Wonders of the Yellowstone first called public attention thereto, was appointed Superintendent of the Park.

INDEX

L

M

O

P

Q

R

S

T

U

V

W

Y